D0872761

SERVICE
ENGINE
SOON
F
E
FUEL DOOR

How To Use
Automotive Diagnostic Scanners

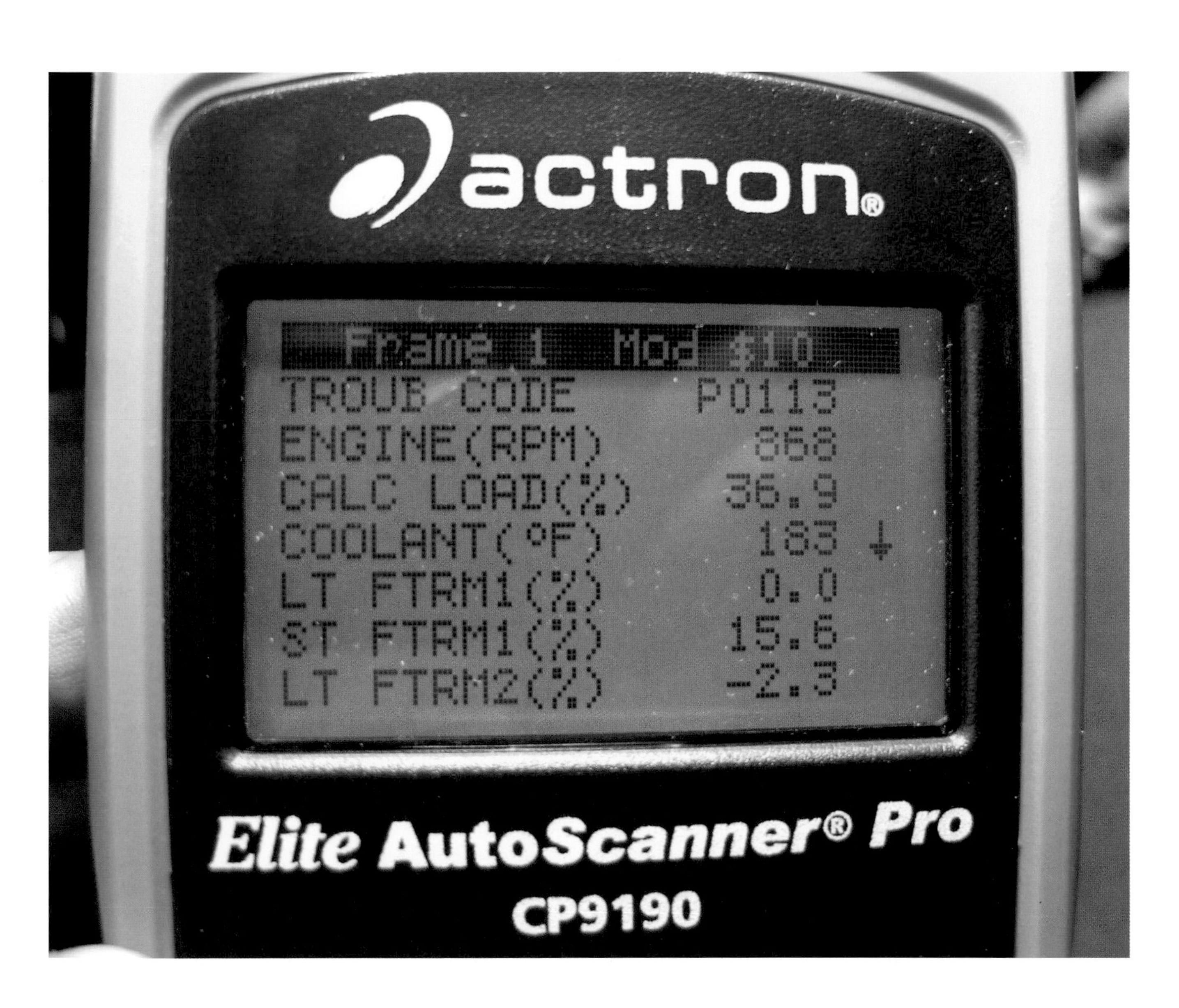

TRACY MARTIN

MOTORBOOKS

First published in 2007 by Motorbooks, an imprint of MBI Publishing Company LLC, Galtier Plaza, Suite 200, 380 Jackson Street, St. Paul, MN 55101 USA

Library of Congress Cataloging-in-Publication Data

Martin, Tracy.
How to use automotive diagnostic scanners / By Tracy Martin.
p. cm.
ISBN-13: 978-0-7603-2800-2 (softbound)
ISBN-10: 0-7603-2800-5 (softbound)
1. Automobiles--Maintenance and repair. 2. Automobiles—Pollution control devices. I. Title.
TL152.M276 2008
629.28'72—dc22

2007029379

On the cover: Automotive diagnostic scanners can be used to isolate problems – to a point. They can lead you in the right direction, but you'll still need to do some automotive detective work. Here, the author and his son take the first step toward repairing their car.

On the frontispiece: A "service engine soon" light is the only way an OBD-II system can communicate with a driver about a malfunction occurring within a vehicle's emission system. Unfortunately, when a malfunction occurs, and the "service engine soon" light turns on, the vehicle owner usually wants the light turned off as soon as possible and is not particularly interested in repairing the vehicle, especially if the repairs might cost a lot of money. While this scenario was frequently feasible with older OBD-I system vehicles, it is much more difficult for automobile owners today (given the greater level of diagnostic capabilities of the newer OBD-II systems) to have this predilection for ignoring warning lights as a viable option.

On the title pages: This Actron scan tool is displaying freeze-frame data. Freeze-frame data is useful when diagnosing why a code was set in the first place, since specific operating conditions are registered and stored at the exact moment a malfunction occurred. *Courtesy Actron*

On the back cover: Whenever a scan tool or code reader is first purchased, it's a good idea to practice using it. This exercise will verify that both the scan tool and OBD-II system are operating as they should. *Courtesy AutoXray*

Editors: Jennifer Bennett & James Michels
Designer: Kou Lor

Printed in China

CONTENTS

PREFACE

When Motorbooks asked me to write this book I did not perceive an immediate need for what I thought would simply be another book on automotive scanners. After all, onboard diagnostics, second generation (OBD-II), has been around since 1996 on every car and light truck sold in the United States.

I did some investigation on what books were available on the subject by starting with several of my 22-year-old son's friends who work on their cars as a hobby. I asked them if I could take a look at the books I assumed they had acquired on scan tools and code readers. I received blank stares in response to the question, but one of them said he owned a code reader. He told me that they used the code reader mostly to turn off the check-engine light when they inadvertently caused it to come on by disconnecting an engine management computer sensor and forgetting to plug it back in when they next drove their cars. I told the group that even a code reader could do more than turn off the check-engine light, and asked them about OBD-II inspection and maintenance monitors and some other basic OBD-II-related questions—again, blank stares. They told me that the code reader came with a 10-page manual, but they did not have any other information.

After a trip to the local bookstore I found nothing on scan tools, code readers, or OBD-II systems, so I tried the Internet—many would agree that if "It" is not on the Internet, "It" probably doesn't exist. Surfing around on my laptop, I did find several books on the subject of OBD-II, but almost all were written for professional technicians, and the few that were not only seemed to cover automotive diagnostics in general and had nothing on scanners and code readers. However, I did find the reason that this book should be written—there are no books that cover the subject of scan tools and code readers for the do-it-yourself technician.

With the availability of code readers and scan tools targeted at the consumer market through retailers such as Sears, Wal-Mart, and auto parts stores, it's more than evident that the aftermarket automotive electronic equipment manufacturers have realized a need for owners and enthusiasts to have access to what once was solely the domain of dealership and professional technicians—an automobile's onboard diagnostic system. What seemed to be missing was a source of information that tied everything together. I wrote this book about scan tools and code readers in the same easy-to-read style as my first two books on automotive and motorcycle electrical systems to fill this information gap.

In this book, the first generation of onboard diagnostics (OBD-I) will be discussed in Chapter 1. Chapter 2 will cover OBD-II, the diagnostic monitoring system in all vehicles sold in the United States since 1996, and the system that code readers and scan tools interface with. Also included is a brief history of automobile air pollution and how this problem has driven the automotive industry to produce these systems in the first place. Chapter 3 covers electronic fuel injection, oxygen sensors, and catalytic converter operation. Code readers are discussed in Chapter 4 with scan tools following in Chapter 5. How an engine works, and especially how to separate engine mechanical problems from OBD-II system diagnostics, is discussed in Chapter 6, and Chapter 7 provides some practical applications for using a scan tool to diagnose emission-related problems.

This book will provide the reader with a sound understanding of how onboard diagnostics relate to engine performance and emission problems. However, because OBD-I and OBD-II systems, and onboard computers—and their numerous sensors and components—are electrical in nature, a basic understanding of automotive electricity will go a long way toward diagnosing and repairing problems with the vehicles that use these systems. My book *How to Diagnose and Repair Automotive Electrical Systems*, also published by Motorbooks, is the perfect companion book to this one. I've also written on the same subject for motorcycles, *Motorcycle Electrical Systems Troubleshooting and Repair*, also published by Motorbooks. You can find more information about these books and some background on myself on my website at: www.tracyAmartin.com. Send me an e-mail if you want to comment on any of the books I have written or just to say hello.

MOTORBOOKS WORKSHOP

How To Diagnose and Repair **Automotive Electrical Systems**

- *Electrical Principles Explained in Plain Language*
- *Choosing and Using Test and Repair Tools*
- *Understanding Ignition, Charging and Control Systems*
- *Reading Wiring Diagrams*
- *Troubleshoot All of Your Vehicle's Electronics*

Tracy Martin

MOTORBOOKS WORKSHOP

Motorcycle Electrical Systems
Troubleshooting and Repair

- *Understanding Electrical Principles*
- *Choosing and Using Tests and Tools*
- *Reading Wiring Diagrams*
- *Testing Ignition and Fuel Injection*
- *Adding Electronic Accessories*

Tracy Martin

I would like to thank the following individuals for helping me with research and information for this book. Without their assistance, I would be lost more than I usually am. Curt Moore and Craig Healy from the SCM Hotline; their technical editing and suggestions saved me from writing something stupid. Paul McCarty, service director for Younger Toyota in Hagerstown, who generously gave me access to the dealership to take photos. Larry Keplinger, who also allowed me to take photos of his exceptional repair facility; Jennifer Grabowski and Bart Ivic from SPX, who helped with technical editing and information on scan tools and code readers; and Elwoods Auto Exchange, where I was able to take many of the photos used in the book. Jen Bennett, my editor at Motorbooks International, and my wife, Leslie, whose editing skills have always vastly improved what I write.

So take a break from working under the hood of your car, sit back, relax, and read all about how scan tools and code readers work with your car or truck's OBD-II onboard computer. Hopefully you'll find what this book contains is entertaining as well as informative.

—Tracy Martin

CHAPTER 1

ONBOARD DIAGNOSTICS, A BRIEF HISTORY

Not a pretty sight of downtown Los Angeles in 1948, as smog obscures the view down this city street. The term "smog" is borrowed from the British, who originated the use of the word in 1905 as a contraction of the words "smoke" and "fog." The first officially recognized "gas" attack (of smog) happened in Los Angeles in 1943. *Photo Courtesy of UCLA Library Department of Special Collections, Los Angeles Times Collection*

Why We Have Scanners, Code Readers, and OnBoard Diagnostics in the First Place

What exactly are automotive scanners, and why do we need them anyway? For years, it seemed, vehicles, vehicle owners, and mechanics got along quite well without them. Where did the need for these tools arise, and do they really do anything? More importantly, what jobs do they perform and are they really necessary when repairing an automobile today? If they are now an essential component of vehicle diagnostics (and they are!), how is the automotive do-it-yourself technician—or even automobile owners with only a passing interest in why the "check engine" light is on—supposed to understand or even read the computer trouble codes and data streams these diagnostic tools produce? These questions, and others like them, frequently leave automotive hobbyists, and even some professional repair technicians, perplexed and without a clue as to the right answers or what direction to take when it comes to high-tech auto repairs.

When discussing automotive scanners, code readers, and onboard diagnostic systems, it is important to provide some background and a little history about the birth and development of these electronic devices before embarking on an exploration of how they operate and what they do in a practical, hands-on manner, and how to use them to make repairs. Let's start with the basics—a brief description of scanners, code readers, and the vehicle diagnostic computer systems with which they interface.

Both scanners and code readers allow a user to receive and view information from a vehicle's onboard engine management computer system. The difference between code readers and scanners is one of quantitative capability: code readers are very limited in the automotive diagnostic information they can provide, while scanners can provide the same information as a code reader, but can also provide additional diagnostic information as well as perform functional testing. By contrast, onboard diagnostic engine management systems perform a number of tasks, including managing fuel injection and ignition systems, shifting automatic transmissions, managing climate control systems, and controlling vehicle security, navigation, communication, lighting, and other computer-related systems. However, by far the most important function onboard computer systems perform in conjunction with the code readers and scanners that work with them (and why these tools are the focus of this book) is to monitor the performance of emission controls, components, and systems, and make the driver aware of when vehicle exhaust is polluting the air.

Scanners and code readers are technically only capable of reading the information onboard vehicle engine management computer systems generate. The onboard computer systems themselves actually monitor all of the engine emissions controls and systems during vehicle operation. Complicating things a bit is the fact that two generations of onboard computer systems exist—known as OBD-I and OBD-II. Originally, onboard computer systems were designed into vehicles by various automobile manufacturers.

These first-generation **o**n**b**oard **d**iagnostics (OBD-I) were developed in the early 1980s and were an attempt by vehicle manufacturers to provide a system that warned a driver/owner whenever there was a malfunction in the emissions' control system. Originally, OBD-I systems and their handheld computer interfaces were designed for use by professional technicians, and each operated uniquely. The information and tests OBD-I systems provided was not standardized among auto manufacturers, and frequently even varied within a single automaker's model years or engine families.

The majority of the first wave of automotive scanners ever produced were manufactured in the United States around 1980. As originally designed, 1980s scan tools for retrieving basic diagnostic information from OBD-I systems used various cables and adapters to plug into the myriad data connectors found on automobiles that were often specific to vehicle year, make, and model. This complexity made these tools expensive to own—many costing thousands of dollars. In addition, they were designed for use only by professional automotive technicians. As a result, because of the cost and difficulty of use, consumers were largely unaware of their existence. In fact, many car and light truck owners at the time (and subsequently, for years to come) did not even know their vehicles were equipped with an onboard computer.

Around 1989, the first code readers were sold in automotive parts stores, finally enabling consumers to tap into some of the information their automobiles had been generating and using for almost a decade. However, it wasn't until 1996 when the automotive industry's exclusivity over vehicle OBD changed significantly: stricter federal emissions' regulations led to standardization of OBD systems across manufacturers. Thus, generation two of onboard diagnostics (OBD-II) was born and standardized, enabling aftermarket scanners and code readers to read any 1996 or later vehicle's onboard computer information. As more and more consumers purchased these tools and demand increased, the price naturally dropped. Today, the average cost range for code readers is between $100 and $200, and for scan tools from $200 to $800.

We will cover the details about OBD-II systems in significantly greater depth as we continue with the remainder of this book, as the primary focus of this book is on modern OBD-II computer diagnostic systems in use today. However, before we continue we will provide a brief overview of the development of scanners, code readers, and OBD-I and OBD-II systems.

It is appropriate at this point to provide a brief history lesson as well, as it will prove useful to understanding how automotive onboard computer systems, and the scan tools and code readers they interface with, came into being, and how and why they developed as they did. In order to clearly understand the evolution and development of diagnostic scan tools, it is useful to start in the 1980s and work backward in time for a bit.

All automotive scanners, code readers, and OBD-I and OBD-II systems were gradually developed for broad consumer use as a direct result of auto emissions' problems from the past. Scan tools, like so much other 1980s automotive and related technology, including electronic carburetors and fuel injection systems, only came into being as a result of auto manufacturers being forced by Congress to clean up the exhaust emissions bellowing from America's tailpipes.

Manufacturers' initial efforts to control auto pollution followed a "Band-Aid" approach, which proved to be unpredictable and unreliable, and in many cases, made the cars and trucks equipped with them undrivable as well. Manufacturers simply did not have compelling economic impetus or significant legislative arm-twisting to force them to develop the engineering technology to control automotive emissions in an effective or standardized manner. As a result, and by way of example, many carburetor-equipped cars from the 1970s would simply stall out at idle when engine temperature got too hot, or the engine would surge at part-throttle because of lean (lack of fuel) carburetor settings that were required to meet emission standards of the day. After much reluctant trial-and-error engineering ingenuity, auto manufacturers discovered the only consistent and reliable means to effectively reduce automotive tailpipe emissions was to utilize computer systems and related technology that could address and deal with all the variables of engine performance. Once automotive engineers discovered and confirmed the viability and attractiveness of onboard computer systems as a means of controlling vehicular emissions, a new set of problems emerged. The new problems were not anticipated, as they dealt primarily with an inherent lack of communication with, and understanding about, the vehicle's onboard computer by the owner/driver or automotive technician.

With the introduction of automotive onboard computers, technicians had to have a means of communicating with these devices. Early computer systems used a check-engine light that simply blinked on and off or in more sophisticated models, the onboard computer used the light to flash out diagnostic trouble codes (code numbers assigned by manufacturers to specific malfunctions in the emissions control system). With the necessary skills, a trained technician could read the trouble codes based on the sequence displayed by the flashing light on the instrument panel.

Initially, the only computer scan tools available to interface with a vehicle's onboard computer system were brand-specific tools, which automakers provided exclusively to their own dealership network. This was a great marketing tool—only new-car dealerships were able to repair whatever went wrong with emission controls systems on their brand of cars and trucks. Fortunately for the automotive aftermarket—and eventually for the rest of us—Congress declared this monopolistic practice illegal.

In the aftermath of the congressional legislation, several electronic tool manufacturers introduced professional-grade scanners in the early 1980s designed for use by independent repair shops. Today, with the ever-growing number of do-it-yourself technicians working under the hoods of their own vehicles, the availability of inexpensive scanners and code readers provides automobile owners with the freedom to choose to not be dependent upon a repair shop or automotive dealership to get their check-engine light to turn off, or to read and understand the diagnostic trouble codes that their vehicle's onboard computer generates.

However, long before the commonplace availability of scanners, code readers, and onboard diagnostic systems, there was smog. As we shall see, smog has played an integral part in the need for, and mandatory development and widespread use of, these tools.

In 1952, protesters wearing World War II gas masks in Pasadena, California, took a stand to register their discontent about the state of the air quality at the time, or lack thereof. I can personally recall in 1959 and the late 1950s having recesses, and sometimes entire school days, cut short on many hot Southern California days in Pasadena due to smog. I remember clearly that it was often difficult to take a deep breath and my eyes would sting from thick brown smog. Oftentimes, the view of the San Gabriel Mountains (less than a few miles from my elementary school in Pasadena) was completely obscured by smog, and the sun would cast a reddish-brown glow. *Photo Courtesy of UCLA Library Department of Special Collection, Los Angeles Times Collection*

Air Pollution—A Historical Perspective

As briefly mentioned, the need for onboard diagnostics, scanners, and code readers came into being due to a dramatic increase in the number of vehicles on the road, starting in the late 1940s, which inevitably led to an increasing amount of automotive emissions, which, unfortunately, led in a direct and unstoppable chain of cause-and-effect to the all-too-familiar problem of air pollution (and most educated people would argue, subsequent global warming). As a result, two generations of onboard diagnostics (OBD-I and OBD-II) exist, along with automotive scanners and code readers that communicate with these systems. Consequently, how all of these developments relate to, and evolved from, our interaction with vehicles and the air we breathe, is worth a closer look.

In the summer of 1943, while the United States waged war in Europe and Asia, Los Angeles experienced what it officially recognized for the first time as an attack of extreme air pollution, which, borrowing on the term originally coined by the British, was termed *smog*. According to the *Los Angeles Times*: "A pall of smoke and fumes descended on downtown, cutting visibility to three blocks." Striking in the midst of a heat wave, the "gas attack" was nearly unbearable, gripping workers and residents with an eye-stinging sensation and leaving them suffering with respiratory discomfort, nausea, and vomiting.

The day after the smog attack, the local municipal government blamed Southern California Gas Company's Aliso Street plant, for its manufacture of butadiene, an ingredient found in synthetic rubber. The plant was temporarily closed for several months, but in the following years, the problem persisted, even after the company spent $1.5 million (a lot of money in those days) to eliminate nearly all of its chemical fumes by completely enclosing the manufacturing process. What local politicians failed to mention, or apparently to thoroughly investigate, was the fact that Los Angeles had had problems with air pollution long before 1943. In fact, as early as 1903, city records reveal that industrial smoke and fumes were so thick that many residents mistook the conditions for a solar eclipse.

However, it wasn't until 1952 that a link between smog and vehicle emissions was officially confirmed. That year, commercial farmers located near Southland refineries complained of unusual crop damage. The leaves of orange trees, an important California agricultural crop at the time, were discoloring or bleaching—a phenomenon not seen in other parts of the country. Furthermore, tire manufacturers disclosed that rubber was apparently deteriorating faster in Los Angeles than in other areas of the country. Spurred on by these unusual circumstances, Dr. Arie Haagen-Smit, a chemistry professor at California Institute of Technology (Caltech) in Pasadena, was urged to investigate the underlying source of these problems. Dr. Haagen-Smit was the first to determine the primary ingredient in smog, now commonly referred to as ozone, was not an ingredient in, or a direct end product of, tailpipe or smokestack emissions, but was, in fact, created in the earth's atmosphere. He discovered that when atmospheric conditions were right, sunlight would act as a catalyst in a photochemical reaction that combined the hydrocarbons from oil refineries with the partially unburned fuel contained in exhaust from automobiles with nitrogen oxides to form ozone (smog). Researchers at Caltech were able to show that rubber exposed to high ozone levels could develop cracks in just seven minutes. This was such a reliable phenomenon that early methods for measuring ozone levels included the highly scientific act of stretching rubber bands around jars and then timing how long the bands took to snap. Obviously something had to be done.

THOUSANDS DIE FROM EFFECTS OF "KILLER" SMOG IN LONDON

The word *smog* is first recorded in 1905 in a newspaper account of a meeting of a British governmental health agency during which Dr. Harold Antoine des Voeux submitted a paper entitled "Fog and Smoke," in which, in the words of the *Daily Graphic* of July 26, "*. . . it required no science to see that there was something produced in great cities which was not found in the country, and that was smoky fog, or what was known as "smog."*" The next day the British newspaper, *Globe,* also commented: "*Dr. des Vœux did a public service in coining a new word for the London fog.*" However, this was not to be the only time air pollution would be officially noted and ascribed as a serious health problem in the United Kingdom—in fact, far from it. In Glasgow, Scotland, winter inversions of the atmosphere and smoke accumulations from burning coal killed 1,000 people in 1909.

More notably, in December 1952, a toxic mix of dense fog and sooty black coal smoke killed thousands of people in London. Smoke pouring out of London's chimneys mixed with natural fog and cold weather conditions caused dense smog to accumulate. The cold temperatures in turn caused Londoners to heap more coal on their fires, making more smoke and smog. The vicious cycle eventually caused catastrophic results. Eyewitnesses likened the killer fog to "a load of car tires on fire." On December 5, visibility was down to 50 feet within minutes. By December 6, 500 people were dead. By December 7, visibility was less than a foot. Ambulances stopped running and gasping Londoners had to struggle as they walked through the smog to city hospitals. By the time the wind blew the toxic cloud away, thousands were dead. In fact, according to a recent study in Environmental Health Perspectives, "as many as 12,000 people may have been killed by the great smog of 1952."

The lethal smog attack in London in 1952 remains the single deadliest environmental episode in recorded global history. Prior to this event, people who lived in large cities more or less accepted dirty air pollution as part of city living. In the aftermath of this incident, many worldwide governments seriously questioned, and attempted to eradicate or at least limit, the poisonous side effects of the industrial age.

Smog is now classified into three types: London smog, photochemical smog, and smog from burning biomass. As described, London smog arises from the mixture of the natural atmosphere with the by-products of coal used to heat homes and businesses. During cool, damp periods (typically in the winter), coal soot and sulfur oxides can combine with fog droplets to form a dark acidic fog. Fortunately, London smog is largely a phenomenon of the past, as most modern heating sources in Europe and the United States use cleaner-burning fossil fuels, such as oil and natural gas. Also, the use of alternative energy sources, like hydroelectric and nuclear energy, have also contributed to the elimination of London smog.

Unfortunately, the other two forms of smog are still with us—one of which, in fact, is significantly worsening and contributing to serious global environmental concerns, like global warming. Photochemical smog, the most prevalent form of smog, is more of a haze than a true fog. It is produced by chemical reactions in the atmosphere triggered by sunlight. A combination of volatile organic compounds (hydrocarbon exhaust pipe emissions) and nitrogen oxides (NO_x) produce an oxidant—ozone—along with other irritating chemicals that combine to produce photochemical smog. The other type of smog is the oldest type of smog known to man. It is produced from the burning of wood. This type of smog combines aspects of both London smog and photochemical smog, since the burning of wood, or biomass, produces large quantities of smoke, as well as other volatile organic compounds (VOC) and NO_x.

Early Automotive Emissions Regulations

In the 1950s and 1960s, the problem of air pollution continued to worsen in Southern California, motivating state and local governments to conduct studies to determine the potential sources of smog. Ultimately, these studies confirmed the overwhelming bulk of smog was attributable to automobile emissions. With a vehicle population of eight million at the time, and billions of miles being driven annually, the Motor Vehicle Pollution Control Board (MVPCB) was created in 1960 to regulate automotive emissions. The first pollution control device was mandated in 1966—a requirement for a positive crankcase ventilation (PCV) valve to be equipped on all vehicles sold in California. Using engine vacuum, a PCV valve sucked up unburned fuel created by combustion gases escaping past piston rings into the crankcase. The unburned fuel was then returned to the intake manifold, where it was burned, instead of being allowed to simply vent into the atmosphere. Obviously, vehicle manufacturers were resistant—to say the least—about having to add what they considered an unnecessary component onto the engine of every vehicle, but economic reality dictated that they had no choice if they wanted to sell automobiles to California residents. The effectiveness of this device remains evident today, as it is still in use in many cars and light trucks.

The California Air Resources Board (CARB) was formed in 1967. Its established purpose was to oversee and regulate air pollution in the entire state. CARB developed testing procedures and regulations that were eventually adopted by Congress in the form of federal emissions legislation. Congress passed the Clean Air Act in 1970 and simultaneously created the U.S. Environmental Protection Agency (EPA). The Clean Air Act called for a 90 percent reduction in motor vehicle exhaust emissions by 1975.

Naturally, serious resistance ensued from automobile manufacturers, since the new mandatory technology was anticipated to cost millions of dollars to implement. Dire predictions about doubling—at a minimum—the manufacturing cost and end consumer price of new cars and trucks were rampant, but in the end, despite considerable opposition, auto manufacturers were forced to comply and did (even though for some, compliance wasn't achieved until 1981).

As a result of this federal legislation, the reduction of three major exhaust pollutants, carbon monoxide (CO), hydrocarbons (HC), and nitrogen oxides (NO_x), was accomplished by implementing a series of emissions controls, including exhaust gas recalculation (EGR), charcoal canister vapor recovery systems, and three-way catalytic converters. Up until the point of the federal legislation, the majority of emission

Still the same after all these years, a positive crankcase ventilation (PCV) valve is a simple device that regulates the flow of hydrocarbons from an engine's crankcase by forcing them into the intake manifold, where instead of polluting the atmosphere, the leftover fuel is burned in the engine's combustion chamber. *Courtesy Kiplinger's Automotive Center, Hagerstown, Maryland*

control measures and systems operated independently from each other, and were controlled by mechanical means. It wasn't until 1981, when the introduction of onboard computers, three-way catalytic converters, and oxygen sensors occurred, that engine performance and overall drivability of vehicles that used emission controls improved.

So what exactly does smog, and the related legislative efforts to minimize it, have to do with the development of automotive scanners and code readers? Everything. The existence and development of two generations of onboard diagnostic systems and the creation and use of automotive scanners and code readers to interface with automotive computers, are integrally linked with the effort to clean up the air we breathe and have unfortunately polluted. That first "gas attack" in Los Angeles in the summer of 1943 was the official start of a war on smog that has now been going on for well over a half-century. What started out simply, with voluntary emissions' regulations and bans on the burning of trash in backyards, progressed to federal legislative requirements being imposed upon all automobile manufacturers requiring all vehicles to have the capability to monitor their own emissions while operational and to warn the driver about any failures of emission controls. Now let's take a closer look at how OBD-I systems were developed and how they operate.

This Ford EGR valve adds exhaust gases back into the engine's intake manifold, which lowers part-throttle combustion temperatures. This in turn reduces NO_x emissions from the exhaust. Engine vacuum operates the valve, and the plastic sensor on top measures valve position (how far open or closed it is), and relays this information to the onboard computer.

Onboard Diagnostics—Generation One or OBD-I

All internal combustion engines produce exhaust emissions as a result of incomplete combustion of the air/fuel mixture, which is caused by the absence of sufficient amounts of available oxygen to completely burn all the fuel present during the combustion process. Because the amount of unburned fuel is so small, fuel economy is not typically an issue. However, when a reduction in emissions was mandated by the federal government, the small amount of unburned fuel became problematic and of paramount concern. Three organizations, CARB, the EPA, and the Society of Automotive Engineers (SAE), started serious research on this issue in 1980, and by 1988, the first generation of computer onboard diagnostic systems (OBD-I) were required to be installed on every vehicle sold in California, with the rest of the nation soon to follow in its footsteps.

To effectively reduce toxic emissions, automakers had to come up with a computerized emission control system

This OBD-II catalytic converter is quite a bit smaller than the ones used on OBD-I vehicles. Unlike OBD-I converters, the OBD-II converter is placed right next to the exhaust manifold, which shortens the time it takes for the catalytic converter to reach operating temperatures. The smaller size also provides a desirable weight reduction over older designs. *Courtesy of Younger Toyota, Hagerstown, Maryland*

This early 1980s General Motors OBD-I-type electronic control module (ECM) had more computing power than the computer used to land astronauts on the moon. An electronically controlled carburetor uses an ECM like this one for fuel delivery control. This computer also controlled other emission system components and the ignition system.

that could perform the following functions: (1) Respond instantly to supply the exact air/fuel mixture for any given driving condition; (2) calculate the optimal time to fire an engine's spark plugs for maximum engine efficiency; and (3) perform both of these tasks without adversely affecting engine performance or fuel economy.

Up until the Federal Clean Air Act deadline of 1975, fuel delivery, at least for the vast majority of vehicles, was primarily accomplished by a carburetor, while ignition timing was determined by mechanical means—i.e., an ignition distributor using springs, weights, and engine vacuum. These old mechanical systems had been in use for more than 80 years, and consequently, were not precise enough or fast enough to meet the new, stricter emissions standards. Basically, a plan was needed that called for a carburetor and ignition distributor with brains. The introduction of the automotive onboard computer provided the means to accomplish the task; the OBD-I system was designed to, and was supposed to, effectively monitor its own performance with regard to tailpipe emissions.

Though not very sophisticated by today's standards, the automotive computers used in the early 1980s were fast enough and accurate enough to effectively control fuel delivery via an electro-mechanical carburetor or electronic fuel injection system. Computer-controlled systems utilize software programs with specific preset reference values for air/fuel ratios and spark advance (ignition timing). By monitoring inputs from various sensors—including engine temperature, engine speed, air temperature, engine load, road speed, transmission gear selection, exhaust gas oxygen, and throttle position—and then comparing the resultant values against an internal reference library, the computer is able to make incremental corrective adjustments hundreds of times each second to maximize the air/fuel ratio to optimal levels. By commanding the output devices under its control, including fuel delivery solenoids (carburetors), fuel injectors, ignition modules, EGR valves, and idle speed controls, the computer is able to keep the engine operating within proper optimal preset values, and in the process, keep emissions at acceptable levels.

OBD-I was originally designed to monitor the performance of fuel delivery systems, at least as related to emissions output, as a means of warning the driver if something went wrong that might potentially cause an increase in pollution. Additionally, OBD-I was theoretically designed with the ability to self-diagnose fuel and ignition systems, and to monitor their ability to keep emissions under control. The OBD-I system's ability to display trouble codes that would essentially instruct service technicians which part or parts to replace if something went wrong was viewed as a side benefit by auto manufacturers.

By the early 1980s, most automotive engineers were of the opinion that auto mechanics of the near future would not need to have a high degree of technical training—or need an in-depth understanding of onboard computer technology—in order to successfully repair automobiles. It was believed that a vehicle's onboard computer and OBD-I system were sufficient to determine a fault in the fuel or ignition systems, and that it would set and display the appropriate trouble code, thereby enabling the mechanic to simply read the code by watching a flashing light on the dashboard and replace the specified malfunctioning component.

However, as things turned out, engineers and automakers were in for a big surprise. Part replacement, as dictated by trouble codes, ultimately didn't actually fix many of the disabled or malfunctioning vehicles. In fact, poor connections in computer wiring harnesses, engine vacuum leaks, bad computers that couldn't diagnose themselves, bad sensors that wouldn't set a trouble code, and engines with mechanical problems all played havoc with the auto repair industry for several years. Instead of unskilled mechanics with little training using the OBD-I system to diagnose and repair problems, just the opposite happened: automotive repair technicians had to have a higher level of training than ever before. The simple process of reading and understanding OBD-I data streams was not really simple at all, since there was no, or very little, information available to independent automotive technicians as to how to do this. Furthermore, this newfound lack of expertise was also partly due to the fact that automotive systems, which were previously unrelated, such as the ignition, fuel, and exhaust systems or emissions controls, were now all electrically connected together via the onboard computer. If something went wrong with one system, a trouble code might be set to indicate a problem existing in another system. Erroneously relying on the information from the vehicle's computer, which they believed they understood, technicians all too often went off on a wild-goose chase, spending hours looking for a problem that existed in an unrelated area of the engine's various systems and controls. These frustrating exercises in diagnostic futility were costly and time-consuming, and stood out in stark contrast to the simpler times of automotive repair before the advent of computer-controlled diagnostics.

For example, in the good old days—before automotive computers—if a car arrived at an auto repair shop with black smoke pouring from its exhaust pipe, the mechanic on duty knew he would soon be repairing or rebuilding a carburetor—he didn't even have to open the hood to figure this out. However, when a carbureted car equipped with an OBD-I system came in with black smoke coming from the exhaust, diagnosis wasn't as easy, because the problem could be the carburetor (which now had wires connected to it), but it could also be the computer itself or one of its sensors, or it could be a bad or poor electrical connection, a vacuum leak, or even an exhaust leak. Adding to the diagnostic dilemma, there may or may not have been a trouble code set and the check-engine light on the dashboard may or may not have been turned on. The mechanic, who now fully understood why his or her title had recently been elevated to technician, had to spend some quality diagnostic time to get to the bottom of the root cause of the problem. Not only did technicians have to have a sound understanding of basic engine operation, they now also had to know how the engine management system controlled the fuel and ignition systems, and how these systems interrelated with various sensors and controls.

Unfortunately, due to a prevailing lack of understanding about how the new computer-controlled systems operated, combined with insufficient training of automotive technicians, many perfectly good components were replaced unnecessarily. Without proper training to deal with the new technology, many in the automotive industry, including technicians and independent repair businesses alike, choose to simply quit instead. Fortunately, those who stayed were forced to acquire a strong desire to learn how electronic ignition systems, computer-controlled carburetors, and cutting-edge electronic fuel injection systems worked. But most importantly, they had to learn how to communicate with a vehicle's onboard computer.

To make things even more interesting, since there was no standardization required between automakers, each

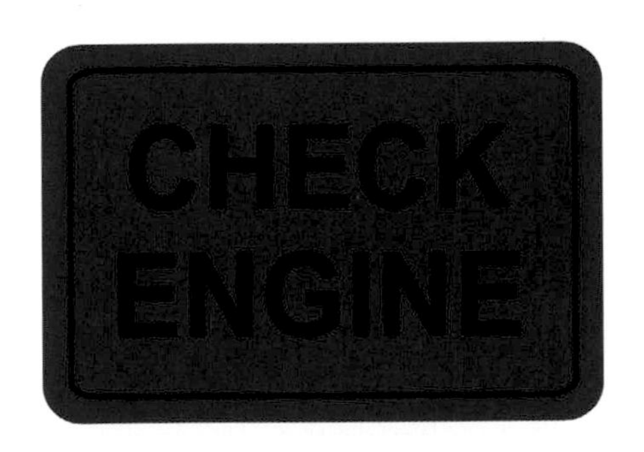

OBD-I check engine lamps were supposed to warn drivers when a vehicle had a malfunction that affected tail pipe emissions. Various means were used to turn them off, including removing the bulb, using an ice pick to break the bulb, and even, on some occasions, fixing the actual problem that caused the light to turn on in the first place.

Because manufacturers were not required to standardize OBD-I diagnostic systems, the Monitor 4000E from OTC needed lots of cables and adapters in order to connect to vehicles from General Motors, Ford, Chrysler, and Toyota. The small square boxes to the right of the scan tool are software cartridges for different years and makes of vehicles. This scanner dates from early 1990. At the time, it was much too expensive for home technicians. *Courtesy OTC Tools*

manufacturer had free rein to design and implement its own OBD-I components and strategies. Consequently, some makes and models of vehicles had more diagnostic information available than others, and some required a scan tool to retrieve trouble codes, while others used a flashing light on the dash to convey codes to technicians. Different methods were required to get onboard computers to cough up their respective stored information. In some instances, certain cars and trucks needed two wires to be jumped together at the vehicle's diagnostic link in order to cause the light on the dash to flash a sequence of codes. Moreover, diagnostic connectors were often difficult to find, since they could be located under the hood, behind dash panels, or tucked away in the center console. In addition, these diagnostic systems came in many different sizes and shapes and required a variety of connectors and cables before they could properly work with various scan tools. The trouble codes themselves were also unique to each manufacturer, and in some cases, differences in, or even different types of, onboard computer systems existed between makes and model years of vehicles from the same manufacturer. Thus, manufacturers had discretion in design engineering as to how "bad" a component would actually have to be before it would cause the check-engine light to turn on, or when it would miraculously turn off if the problem mysteriously fixed itself. Even the computers themselves had different names—depending upon which make of vehicle they were installed into—some of the more common names included Chrysler's Logic and Power Modules, Single Module Engine Controller (SMEC), and Single Board Engine Controller (SBEC); General Motor's Computer Command Control (CCC) and Electronic Control Module (ECM); and Ford Motor Company's Electronic Control Assembly (ECA), Multifunction Control Unit, (MCU), and Electronic Engine Control I, II, III, and IV (EEC-I to IV); along with countless others, including those associated with European and Asian imports. Consequently, consulting a service manual or otherwise having access to vehicle-specific OBD-I information, and then knowing which procedures to perform, was absolutely essential if there was going to be any chance at all at figuring out what was wrong with a system and how to go about making the appropriate repairs.

From an emissions point of view, the first-generation OBD-I systems had some serious drawbacks—many of

Because OBD-I diagnostic systems varied so widely between various car and truck manufacturers, a vehicle-specific service manual and a scan tool were absolutely essential to properly and adequately diagnose check-engine lights and associated drivability problems.

these systems couldn't even determine if an engine and its pollution control systems were actually operating properly. In fact, the vehicle could actually be spitting out lots of raw, unburned fuel from its tailpipe, and the computer, despite its OBD-I monitoring system, would not turn on any warning lights—instead conveying the message to the driver to just keep driving and that everything's OK.

A frequent example of this type of scenario occurred whenever the vehicle's ignition system had an intermittent misfire. In such cases, raw unburned fuel would get sent through the catalytic converter, and in the process, cause high exhaust emissions that could ruin the converter. In an effort to compensate and correct the problem, the onboard computer would try to lean out the fuel system. Unfortunately, in such instances, the computer wasn't really able to fix this type of problem, but as long as all the sensors remained within preset electrical values, the dash warning light or check-engine light would never turn on as it was originally intended to do in just such a scenario.

Another example of the serious shortcomings of the first generation of OBD-I computer systems was with their interaction with catalytic converters. Converters would frequently get plugged up, whether from an overly rich fuel system, coolant leak into the combustion chamber, or just from high mileage, causing a severe lack of engine power. Consequently, the converters were often illegally removed and never replaced. OBD-I systems were unable to detect that the converter was missing, so no check-engine light ever came on, and the vehicle could be driven endlessly, with no apparent emissions problems, at least as far as the poor unaware owner was concerned.

Various versions of OBD-I vehicles were in mass production by U.S. manufacturers as early as 1980, even though OBD-I compliance was technically not required until 1988. Early OBD-I cars and trucks used computer-controlled carburetors at the heart of their fuel systems; these computer-controlled carburetors were later replaced by more sophisticated electronic fuel injection systems. Even though the earliest produced OBD-I vehicle is nearly 18 years old today, it's still important to understand the basics of how each manufacturer interpreted the original legislative OBD-I mandates, as well as to know how to read codes and perform diagnostic testing on these vehicles, since there are a large number of 1980s car and truck enthusiasts. As these vehicles age into the next generation of classic cars, knowing how their computer systems work will be useful to maintaining them in restored condition.

Three of these data link connectors fit OBD-I systems. From right to left are: Chrysler, Ford, and General Motors. These connectors were used from the early 1980s until the arrival of the next generation of OBD-II computer systems in 1996. The OBD-II connector is at the far left. *Courtesy Actron*

ELECTRONIC CARBURETORS

This Rochester Dualjet carburetor used a solenoid to control metering rods to vary the amount of fuel delivered to the engine. The computer-controlled carburetor was an intermediate step between older carbureted technologies and newer electronic fuel injection systems. The black plastic connector, located on the top of the carburetor, controlled the fuel-mixture solenoid, while the white connector (right) was for the throttle position sensor (TPS). The funny-looking thing at the far right is an idle speed control motor.

Most early OBD-I vehicles used an electronically computer-controlled carburetor for fuel delivery. At the time, technicians often speculated that auto manufacturers needed to use up all the carburetor castings and machine tooling before they could make the transition to electronic fuel injection, while automotive engineers argued their employers needed time to determine how best to mass produce reliable fuel injection systems. The truth probably lies somewhere in between. Consequently, computer-controlled carburetors existed in large production numbers from 1980 until around 1988. Electronic fuel injection systems were phased in during the mid-1980s. The last carbureted vehicle sold in the United States was made by Isuzu in 1994.

In the early 1980s, many independent repair shops operated in crisis mode whenever vehicles using carburetors with wires coming out of their castings started showing up with drivability problems. Since most automakers were less than forthcoming when it came to sharing diagnostic and repair information, most mechanics at the time were not prepared to deal with the new automotive computer technology. Not only did mechanics have to understand how a carburetor really worked,

Onboard Diagnostics—Generation Two or OBD-II

OBD-I systems were primarily in use after 1980, and were doing a fairly adequate job of meeting federal emission standards of the day. However, like onboard computer technology, emission regulations continued to evolve and were, and still are, forever increasing the limits upon exactly how much "bad" stuff can come out of a vehicle's exhaust. Naturally, a new standard was on the horizon—the second generation of OBD computer diagnostics, or OBD-II. Automakers were given several years lead-time to develop this new technology. General Motors was the first to use it on a limited basis starting in 1994. By 1996, all cars and light trucks sold in the United States were required to have OBD-II systems.

There are several primary differences between OBD-I and OBD-II systems. The most notable difference is they also had to have a sound understanding of automotive electricity and electronics. Most didn't. As if that wasn't bad enough, they also needed an ability to grasp abstract concepts as they related to computer inputs and outputs. Needless to say, initially there were very few technicians who possessed the unique requisite skill set and knowledge to get the job done. Manufacturers of automotive diagnostic test equipment, such as Sun, Allen, and Bear, provided training programs to remedy this problem—no doubt, in the hopes of selling more equipment. Technicians who were unable to adapt and learn the new technology went on to do something else, instead of diagnosing drivability problems on automobiles. This abrupt change in technology caused many independent repair facilities to permanently close their doors.

Fortunately, the reality was that many of these computer-controlled carbureted systems actually worked pretty well. Unfortunately, a sadder reality was that these new systems were not widely understood and misadjustments and wasted repairs were all too common, resulting in these early systems receiving a negative reputation. It was common practice for uneducated mechanics in unfamiliar territory to disconnect the wires coming from a carburetor as the primary method of solving most drivability problems. Naturally, this solution to the problem made the engine run even worse than when it was first brought into a repair shop, as these systems were all programmed to default to an overly rich fuel mixture, which resulted in the dumping of too much fuel into the engine. As these early carbureted cars and trucks started racking up the miles in the mid- to late 1980s, worn-out carburetors were exchanged for rebuilt units. But, as might now be expected, carburetor remanufacturers were no better at rebuilding computer-controlled carburetors than most mechanics. Consequently, the rebuilt carburetors were often worse than the ones they replaced, since the technicians responsible for the rebuilding of the carburetors often mixed up internal parts between carburetor models. In addition, rebuilt carburetors were very expensive—often costing hundreds of dollars more than the standard nonelectrically controlled carburetors prevalent a few years before.

Nonetheless, computer-controlled carburetors did, in fact, do what they were designed to do—they adjusted fuel delivery. They did this by several means, including vacuum solenoids, switches, stepper motors, and electronic solenoids, that controlled fuel-metering rods. Some of the better performing units included Rochester's Dualjet and Quadrajet carburetors and Motocraft's 7200 Variable Venturi carburetor. The Quadrajet carburetors had been around for years before they were redesigned to perform as electronically controlled carburetors; in the computer-controlled configuration they were actually quite successful and tunable. Dualjets were merely Quadrajets cut in half, with no secondary throttle plates. The more notable Variable Venturi (VV) carburetor by Motocraft was similar to some modern motorcycle carburetors, in that its venturi, an opening that air passes through on its way to the intake manifold, was adjustable and changed in accordance with engine vacuum or load. The VV carburetor had a jet needle and a slide to control air and fuel flow, but only one fuel circuit. By contrast, standard carburetors had a fixed venturi and seven fuel circuits. Consequently, in a standard carburetor, several fuel delivery circuits were needed to match fuel flow with airflow. The VV carburetor offered performance that matched electronic fuel injection of the time; it was used on Ford's Crown Victoria and Grand Marquis cars (can you say police cruiser?). Unfortunately, the VV carburetor was also one of the least understood computer-controlled carburetors of the time, and was subject to misadjustment and gross general mechanical abuse from less-than-knowledgeable mechanics.

Nowadays, electronic fuel injection systems have completely replaced the automotive carburetor, rendering it nearly obsolete, except in those cases where it lives on in classic restored vehicles. The trend toward fuel injection systems is here to stay, with expansion of its use in motorcycles, watercraft, snowmobiles, and now even lawn tractors. Most of the vehicles that once used the early computer-controlled carburetors have gone on to the junkyard, but automotive enthusiasts will always keep some of them alive today.

standardization. All vehicle manufacturers were required to follow the same rules and regulations, which included specifications for the design and installation of a universal data link connector on every vehicle, uniform automotive terminology, and consistent trouble code identifiers and definitions. One significant difference between both systems is the OBD-II system's ability to monitor the effectiveness of catalytic converter operation as the vehicle is being driven. By contrast, OBD-I systems were not even able to determine if the converter had been removed from a vehicle. Another major difference is the level of diagnostics available. Where an OBD-I system may have used only one or two codes for a single sensor, OBD-II systems have as many as four or more codes for each individual computer sensor. In addition, automakers were given the freedom to use individualized "P1" codes, which were allowed to be specific to each brand of automobile. When manufacturer-specific P1 codes are combined with the numerous standardized OBD-II generic codes, there are literally thousands of diagnostic codes to cover nearly all of the systems equipped on today's vehicles.

Retrieving Codes and Testing—OBD-I

In view of the large number of older cars and trucks still on the road today, there are still lots of OBD-I systems in need of a little TLC. While the focus of this book is OBD-II systems and the scanners and code readers that interface with them, the rest of this chapter will cover how to perform diagnostic testing on the first generation of OBD-I systems from General Motors, Ford, and Chrysler, the primary manufacturers of OBD-I vehicles.

Many modern OBD-II scanners and code readers for the do-it-yourself consumer market are not capable of performing on OBD-I vehicles. A few scanners do come with OBD-I data link cables and adaptors and will work on earlier OBD-I vehicles; however, these scan tools are more expensive than those that are only capable of interfacing with OBD-II vehicles.

Fortunately, there are no worries—OBD-I codes can still be read, and some computer testing is performed using other methods and tools. The tools required for retrieving trouble codes from OBD-I-equipped vehicles are simple and inexpensive; they include a 12-volt test light, several jumper wires, and several large paper clips.

If nothing else, by taking a look at this older technology, current technicians will gain a greater appreciation for the usefulness, capability, and efficiency of the modern OBD-II diagnostic systems of today. We'll start out by taking a look at the General Motors OBD-I system and trouble codes, and perform some testing of General Motors's computer diagnostics.

General Motors—OBD-I Systems

Early on, General Motors designed one of the more sophisticated OBD-I systems. For at least a few years afterward, the other big automakers had to play catch-up with their own designs and technology. Remember, General Motors referred to its OBD-I onboard computer as an Electronic Control Module, or ECM; thus, ECM and onboard computer will be used interchangeably in this discussion dealing with first generation General Motors onboard computers.

The General Motors OBD-I system is capable of functioning in four states, or modes of operation: (1) Field service state; (2) back-up state; (3) 10K or special state; and (4) road test or open state. Each of these diagnostic test states test different aspects of the vehicle's computer system. The most common diagnostic mode used on these types of vehicles is the field service state, because it is the easiest to use. In field service mode, trouble codes can be retrieved simply by using an appropriate scan tool or by watching the flashing check-engine or service-engine-soon light. (Please note: Not all General Motors vehicles equipped with OBD-I systems can perform all functions; consequently, the specific vehicle service manual should be consulted when attempting to work on these vehicles. In addition, code definitions may differ from one vehicle to the next, and appropriate diagnostic procedures may vary as well. A service manual or other source of vehicle-specific OBD-I information is absolutely essential in order to successfully perform diagnostic tests, and/or to read trouble codes, on these vehicles.)

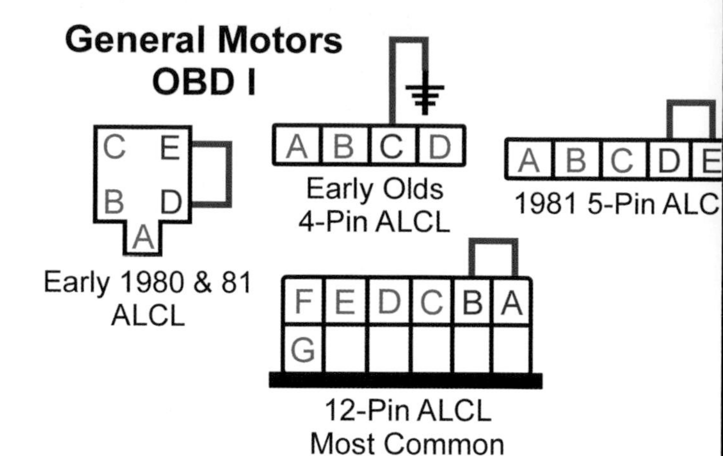

Fig 1-1—General Motors used different assembly line communication link (ALCL) connectors in their OBD-I systems. In some service manuals this connector is referred to as the assembly line data link (ALDL) connector. By far the most common connector in use was the 12-pin ALCL. By placing a jumper wire between terminals A and B on the ALCL connector, the electronic control module (ECM) was triggered into diagnostic mode. Also shown are two 5-pin (top left and top right) and one 4-pin (top center) data link connectors. Jumper wire connections are shown in purple.

Trouble codes are easy to read on General Motors vehicles provided the scan tool used has OBD-I capabilities and a correct General Motors diagnostic cable adapter is utilized. After identifying the year, make, model, and engine type of the vehicle to be diagnosed, simply follow the instructions on the scanner's display to either read trouble codes or erase them from the computer's memory. In

General Motors was one of the few manufacturers that located the early OBD-I ALDL connector next to the steering wheel. It was not until the emergence of OBD-II systems that other automakers followed suit, eventually moving their DLCs inside the passenger compartment within 16 inches of the steering column. *Courtesy Elwoods Auto Exchange, Hagerstown, Maryland*

addition to reading codes, the OBD-I system on General Motors vehicles will display a live data stream on the scanner that can be used to further diagnose engine performance problems. If you don't have an OBD-I capable scan tool, a manual method for retrieving trouble codes (covered more in depth in this section) is easy, once the sequence of the flashing check-engine light is understood.

Retrieving Trouble Codes—General Motors Field Service State

To enable the check-engine light to flash out the trouble codes stored in the vehicle's OBD-I system, it must first be placed into field service state. The first step in this process is to locate the Assembly Line Communication Link (ALCL) or Assembly Line Data Link (ALDL) connector. Generally, the ALDL or ALCL connector is located under the dash on the driver's side of the car. (But not always!) A component locater listed in a service manual can help you properly identify the test connector location. The most common type of ALDL or ALCL connector used has 12 pins (which makes it readily identifiable). The second step in the code retrieval process is to connect pins A and B on the ALDL or ALCL connector together with a paper clip, and turn the ignition key to the RUN position, but do not start the vehicle. The ECM (electronic control module or computer) will then be placed into the field service state (see Fig 1-1).

Once the ignition key is in RUN position, the check-engine light should start to flash out code numbers, starting first with a code No. 12, the default position or baseline. Code 12 is indicated by a single flash (representing the number one or first number in the number 12), followed by a pause, and then followed by two more flashes (representing the number two or second number in the number 12). The code 12 sequence will repeat a total of three times, since all code flashing sequences are repeated three times, in case you miss it. If no malfunctions in the emissions' system are found, no other trouble codes will be displayed, and the default or baseline code 12 will continue to repeat. In essence, a continuously repeating code 12 means the computer system has passed all tests and nothing is wrong (that it can detect).

However, if a malfunction is detected somewhere in the emissions system, a different trouble code will subsequently be displayed by the ECM after code No. 12 finishes flashing three times. Thus, when there is a malfunction, the flashing sequence changes and the assigned malfunction-specific trouble code will be displayed after the baseline code 12 is finished being displayed.

For example, if a malfunctioning TPS sensor (a code 21 as assigned by General Motors) was detected by the ECM, the check-engine light sequence would be displayed as follows: A series of three code 12 pulses would flash, followed by a series of three sets of code 21 flashes. (Code 21 would be displayed as two flashes, a pause, and then a single flash.) If the only malfunction in the emissions system was in the TPS sensor, then after code 21 flashed three times, a series of three repeating baseline code 12 flashes would again repeat. Trouble codes 12 and 21 would each continue to flash out in their respective sequences until the ignition is finally turned off.

General Motors OBD-I systems are capable of displaying two types of codes: hard and soft. A hard code represents a malfunction actually found by the computer in one of the emission control-related systems. A soft code represents a malfunction that occurred sometime previously during prior operation of the vehicle. These soft codes are stored in the ECM's memory, but they do not represent current malfunctions that are actually present in a system at the time the codes are retrieved. Soft codes can be likened to intermittent codes, inasmuch as they have occurred at some interval(s) in the past, but are not currently occurring. A soft code is really nothing more than the computer displaying an old code stored in its memory that was once a hard code. Consequently, a technician viewing a soft code has no way of determining if the soft code represents an old malfunction that has already been repaired and a prior technician simply failed to erase the associated code, or a true intermittent malfunctioning component. Technically, the computer can actually only detect current malfunctions or hard codes and then store them in memory. Thus, for a soft code to truly represent a malfunction that occurs only intermittently, the malfunction must once again show up as a hard code.

To determine if a code is hard or soft, write down all the codes that are flashed out in the field service state (so you can subsequently reference them). Then clear the codes by erasing them from the computer's memory. This can be accomplished by disconnecting the battery (make sure the ignition key is in the OFF position when this is done or the computer can be damaged). Once the codes are erased, repeat the code retrieval process again as previously explained. Any codes that again flash out that are duplicates of the first set of codes will be actual hard codes and will represent malfunctions that are detected by the computer to be currently occurring in various vehicle components. Any trouble codes that were present in the initial code retrieval test but do not show back up in the subsequent code retrieval test represent soft or intermittent codes. Any soft codes found mean that code-specific malfunctions were detected on a previous computer scan of emission control-related systems (and stored in the computer's memory as trouble codes), but are not presently detected to exist by the ECM during the subsequent testing. Thus, the malfunctions originally detected by the ECM are determined to be only intermittent (or potentially resolved) malfunctions that happened only during previous driving condition.

In addition to enabling a technician or other user to read codes, an OBD-I system's field service state will also cause the ECM to energize solenoids and relays. This function is useful for determining if these components are working and are under the control of the computer. Some of the actuators that an ECM is capable of energizing in field service state are the exhaust gas recirculation (EGR) solenoid, the canister purge, and the radiator coolant fan. The idle air control motor will seat, or move into its closed position, on all 1987 and later model port-fuel-injected (PFI) and throttle body-injected (TBI) engines. On certain pre-1987 General Motors vehicles equipped with TBI engines, the ECM in its field service state will cause the idle air control, once energized, to move in and out. Again, the significance of such a test is to determine if the ECM can control these components, and that they are, in fact, working.

Whenever an engine is started in the field service state (leave the paper clip in the ALDL connector), the check-engine light will flash rapidly, at a rate of approximately 2.5 times per second, indicating closed loop operation. Closed loop mode occurs when the ECM and oxygen sensor form an informational loop—the oxygen sensor senses exhaust gas oxygen content and relays this information to the ECM. Based on the relayed information, the ECM corrects the fuel mixture and the process is repeated. By contrast, when the engine goes into open loop operation mode (where the oxygen sensor's input to the ECM is ignored) the check-engine light will flash only once per second. (Additional information on open and closed loop modes of engine operation can be found in Chapter 3.)

In addition, when the check-engine light is on and when it is off is a significant test function as well. In closed loop operational mode, when the light is on, oxygen sensor voltage will register high—above 0.45 volt, indicating a rich air/fuel mixture. When the check-engine light is off, the oxygen sensor voltage will register low—below 0.45 volt, indicating a lean air/fuel mixture. Knowing whether the oxygen sensor voltage is low or high is useful for determining if the oxygen sensor is switching at a proper rate, as it should whenever the vehicle is in closed loop mode, as well as for determining if there are any other malfunctions in the fuel system. (Again, be sure to consult Chapter 3 for additional in-depth discussion of open/closed loop operational modes and oxygen sensor operation.)

General Motors—Back-up Diagnostic State

The General Motors OBD-I system is set to automatically revert to backup state if it has a failure of one of its sensors during vehicle operation. Also known as the limp-in mode, the back-up state allows a computer system that is operating in a less-than-perfect manner to attempt to keep the engine running long enough to get the vehicle to a place where it can be repaired without leaving the driver stranded on the road. While in back-up state, spark advance is fixed, resulting in a lack of engine power, and the effectiveness of the fuel delivery system is determined only from input based on rpm and input from the coolant sensor and throttle position sensor. Any other sensors that are malfunctioning are ignored. In other words, even without all of the computer sensors that are normally required for engine management operational, the ECM will still try to keep the engine running, based on minimal information. The ECM can be placed into backup state by placing a 3,900-Ohm resistor between pins A and B on the ALDL (or ALCL) connector (the same pins used for retrieving trouble codes). On port-fuel-injected cars that cannot be started, placing the ECM into backup state may get the engine started. If so, this usually indicates a faulty mass airflow sensor (MAF) or defective ECM.

General Motors 10k or Special State

The 10k or special state is useful for taking readings of rpm or ignition timing, as this state serves to stabilize the engine. It can make diagnosis easier to understand. It is also useful whenever the need arises to compare one vehicle's readings against another's. Once in 10K or special state, the ECM will bypass its own internal timer and allow the system to go directly into closed loop mode without the customary delay, stabilizing the engine. Ignition timing (spark advance) is also fixed at a predetermined point, further stabilizing the engine, and idle speed is fixed at 1,000 rpm. Thus, a beneficial consequence of fixed spark timing and fixed idle speed is that readings between different vehicles can be compared and interpreted more easily and are more consistent. The ECM can be placed into special state by placing a 10,000-Ohm resistor between pins A and B on the ALDL (or ALCL) connector (the same pins used for retrieving trouble codes).

General Motors—Road Test or Open State

Normal vehicle operation takes place in the road test or open diagnostic state. In this state, there is no need to use a jumper wire on the ALDL. In this state, the ECM performs all normal idle speed and spark advance control functions. If the ignition timing or the idle speed needs to be checked, placing the ECM into road test state will facilitate testing. Most scan tools are capable of automatically reading all ECM data in this state. However, it should be noted that the ECM computers on some older model General Motors vehicles are not capable of out-putting all data information in road test state as this function is not programmed into their software. Consequently, it could appear that the lack of data indicates that something is wrong with the ECM, when nothing is.

This ABS diagnostic link connector was used on Ford Motor Company vehicles from the early 1980s until 1995. This Ford connector is identical to the Ford OBD-I engine management connector, except for color. On Ford vehicles, the OBD-I connector is black or gray and is located somewhere under the hood, at least according to most component locator manuals. *Courtesy Elwoods Auto Exchange, Hagerstown, Maryland*

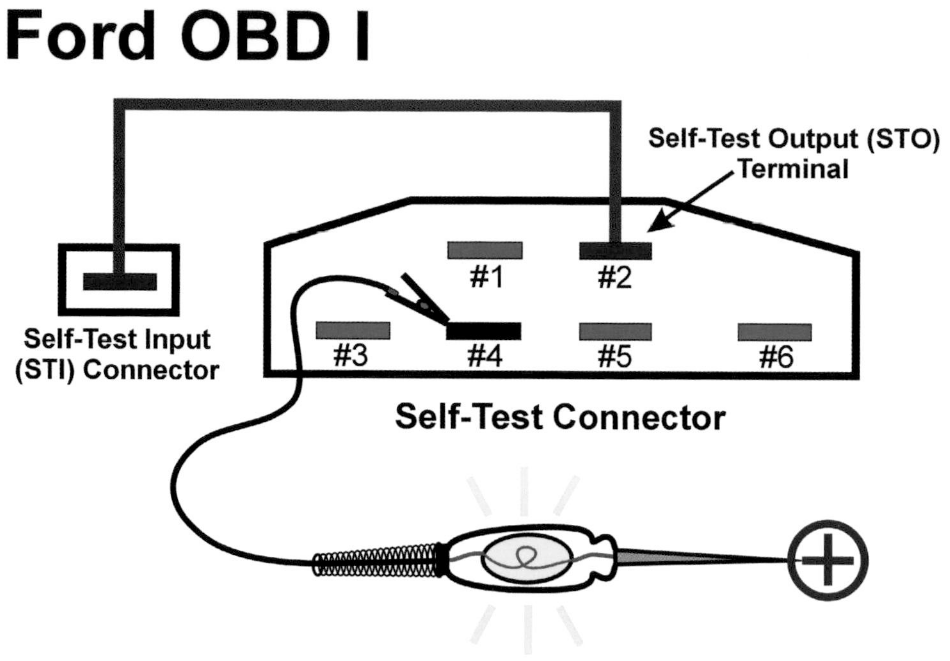

Ford OBD I

Self-Test Output (STO) Terminal

Self-Test Input (STI) Connector

#1 #2 #3 #4 #5 #6

Self-Test Connector

Fig 1-2—Depending on the vehicle, the self-test connector on Ford vehicles is located in different places under the hood, and can be either black or gray in color. A jumper wire connected between terminal 2 (self-test output) and the self-test input (STI) connector will place the electronic control assembly (ECA)—Ford's name for its onboard computer—in self-test mode. A test light connected as shown in the illustration will then flash out the trouble codes. On later model vehicles (after 1986) the check engine light takes the place of a test light for purposes of reading codes.

Ford Motor Company—OBD-I Systems (EEC-I Through IV)

In 1983, Ford Motor Company introduced its fourth-generation Electronic Engine Control (EEC-IV) system. Prior to the release of the EEC-IV, all Ford vehicles were equipped with either EEC-I, II, or III electronic carburetor systems, or microprocessor control unit (MCU) carbureted systems. These early systems were quite limited in their ability to self-diagnose much of anything. Hopefully, most of the cars and trucks that used them are now permanently retired in junkyards. The EEC-IV system and electronic fuel-injected systems were a much-needed improvement over predecessor models, as they could perform several self-tests as well as generate and store trouble codes. A data stream could also be displayed if a scan tool was used as well.

The OBD-I systems on EEC-IV vehicles are capable of performing four basic self-diagnostic tests: (1) Key On/Engine Off self-test (KOEO); (2) Continuous Memory (CM) self-test; (3) Key ON/Engine Running self-test (KOER); and (4) several miscellaneous EEC-IV-related self-tests.

Ford Diagnostic Testing—Key On/Engine Off (KOEO) Self-Test

We'll start with the most common series of tests—the Key On/Engine Off, or KOEO self-tests. The KOEO tests are designed to monitor and/or test various components and circuits controlled by the electronic control assembly (ECA). Prior to commencing any KOEO self-test, the engine must be at normal operating temperature, so be sure to take the car for a short test drive first. To start a self-test, connect a jumper wire between Pin 2 on the Self-Test Connector and the Self-Test Input (STI) connector. A test light is then connected between Pin 4 on the Self-Test Connector and battery positive (see Fig 1-2). The code sequence can be read by either watching the test light flash, or by observing the Service Engine light located on the dash on later model EEC-IV vehicles. (Earlier vehicles may or may not have a check-engine light to perform this function.)

With the test light and jumper wire in place, turn the ignition to the RUN position. The test light will flicker and then flash out the On-Demand codes. On-Demand codes are not stored in the ECA's memory, but rather, they instead represent hard codes (or hard faults) that are occurring contemporaneous with test procedures. On-Demand codes consist of only two digits on vehicles up to 1991, but are three digits on all 1991 to 1995 cars and trucks. On-Demand codes will flash out in sequence from lowest to highest. If no codes are present, a pass 11, or code 111, will flash. After a six-second pause, the ECA will then output a separator code, No. 10 (a single flash). The separator code simply serves to separate the On-Demand codes from other codes to follow. After another six-second pause, the Continuous Memory codes (those codes stored in the ECA's memory) will flash out. These codes represent soft codes (or Soft Faults) or malfunctions that occurred during past vehicle operation but are no longer present during current test procedures. If no Continuous Memory codes are present, a pass 11, or code 111, will again flash.

Once any problems have been repaired, all trouble codes should be erased from the ECA's memory so that some poor technician at a later time doesn't chase after malfunctions associated with what appear to be soft codes that are really not a problem any longer. To erase all trouble codes, start the KOEO self-test, and as soon as the codes commence flash out on the test light, simply unplug the jumper wire between the STI connector and the Self-Test Connector. This will erase all existing codes. Another method that accomplishes code erasing is to simply disconnect the battery for at least 10 seconds. However, it should be noted that disconnecting the battery causes the ECA to lose its adaptive memory, which may cause the engine to run roughly. Consequently, after disconnecting the battery, a "Relearn" procedure should be performed. Consult a vehicle-specific service manual for a detailed explanation on how to perform this procedure, as it is not the same for different years or models of Ford vehicles.

Ford Diagnostic Testing—Output State Self-Test

Once the On-Demand and Continuous Memory codes have been displayed, a self-test in the Output state can be performed. This self-test checks the ECA's ability to control various EEC-IV solenoids and is useful to determine if the ECA and its controlled components are working properly, or at all. To energize or turn on the solenoids, open the throttle and then quickly close it. To deactivate or turn off the solenoids, open and close the throttle again. This is a good test for verifying proper electrical and mechanical operation of the EEC-IV system's solenoids.

Ford Diagnostic Testing—Key On/Engine Running—KOER Self-Test

The Key On/Engine Running (KOER) self-test is similar to the KOEO self-test, only the engine must be running. To perform a KOER self-test, make sure the engine is hot. Connect a test light and jumper to the Self-Test Connector in the same manner as was performed in the KOEO self-test, and then start the engine. The ECA will pulse (flash) an engine ID code—two pulses for a four-cylinder engine, three pulses for a six-cylinder engine, and four pulses for a V-8 engine. Engine ID codes don't have anything to do with the self-test; they are simply confirming the type of engine in use.

After the engine ID code pulses, a pause that lasts between 6 and 20 seconds will follow. On most vehicles a Dynamic Response code 10 (a single flash) will then flash. You now have about 15 seconds to "goose" (Ford's actual terminology) or snap open the throttle. The ECA will monitor any changes in engine rpm, TPS voltage, and MAP frequency in response to the throttle opening, and will determine if anything is amiss. If any of the responses received are not what the ECA expects, a new trouble code will then be pulsed, which will indicate a problem with one of its sensors.

Once the "goose" test is complete, the test light may flicker. After that, there will then be another six-second pause. The On-Demand codes will then flash out next. If there are no On-Demand codes, a pass 11, or code 111, will flash. After the On-Demand codes have finished flashing, a computed timing self-test can be performed next. The ECA will fix the ignition timing for two minutes at 20 degrees (+/-3 degrees) above base ignition timing, allowing ignition timing to then be checked with a timing light for accuracy.

In Ford's Wiggle self-test mode, the ECA will flash the check engine light if there are any opens or shorts in any computer wiring. A screwdriver handle is used to tap on sensors and connectors and the heat gun will cause connectors to expand. Heating or tapping on sensors and connectors is a good way to stress electrical connections and amplify any connector problems.

Ford Diagnostic Testing—The Wiggle Test

A Ford Wiggle Test is a good way to find intermittent electrical problems with a computer system's wiring harness and connectors on Ford OBD-I vehicles. Perform this test only after any Continuous Memory codes have been erased, so no invalid or misleading soft codes will display and only current malfunctions can be detected. To put the ECA into Wiggle Test mode, connect a test light to the self-test connector in the same manner as in the KOEO self-test (vehicles with a check-engine light do not require a test light for monitoring). Turn the ignition to the ON position. Connect a jumper wire between Pin 2 on the Self-Test Connector and the STI connector. Next, remove the jumper and then immediately reconnect it in the same place. If the ECA detects any intermittent shorts, or opens, in any of its sensor circuits, the check-engine light, or test light, will momentarily flash. With the ECA in Wiggle Test mode, the wiring harness can be wiggled and/or the computer sensors can be tapped with a screwdriver handle or heated with a heat gun. Again, if the ECA then detects an electrical short or open, the check-engine light will flash. Just like the KOEO and KOER self-tests, the Wiggle Test can be performed during a test drive.

Ford Diagnostic Testing—Power Balance Test

On any model of Ford vehicle equipped with a sequential fuel injection system, an OBD-I diagnostic system can perform a cylinder Power Balance test. This will test the engine's mechanical condition. After the On-Demand codes have finished pulsing, hold the engine idle speed above 1,000 rpm for two seconds in order to initiate a Power Balance test. Then let the engine return to idle. The ECA will then perform the first level of its Power Balance self-test. During this test, the ECA is looking for the same drop in rpm each time it shuts off a cylinder's injector. Thus, once a cylinder's fuel injector is turned off, the computer reads the rpm drop; the computer then does the same thing to each fuel injector sequentially. If all of the fuel injectors show the same rpm drop when they are shut off, the test is successfully passed and a code 9 (nine flashes) will be pulsed. If any one of the cylinders fails the test, that cylinder's number will instead be flashed—if cylinder three fails the test, the test light will flash three times. If any cylinder fails, the Power Balance self-test can be repeated at the next level of sensitivity by snapping the throttle open. Each level of sensitivity will lower the amount of acceptable rpm drop that the ECA looks at. Again, any cylinder that fails will be indicated by an appropriate series of pulses in the test light. Snapping the throttle open again at a higher (third) level will render the test even more sensitive to the drop in rpm measured by the vehicle's ECA.

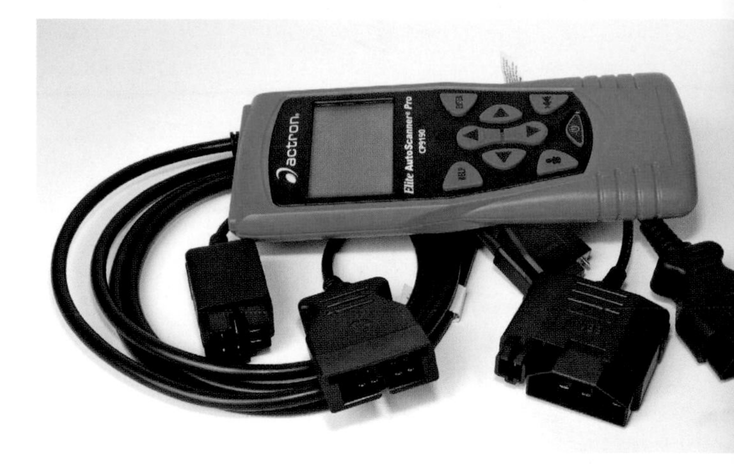

This Actron CP9190 scan tool can interface with most domestic OBD-I vehicles. The OBD-I connectors included with the tool are (left to right) for the following: Chrysler/Jeep, General Motors, and Ford. It is also capable of reading OBD-II and CAN data from any vehicle from 1996 or later using the far right connector. *Courtesy Actron*

Daimler Chrysler Corporation—OBD-I Systems

Chrysler first introduced its version of electronic fuel injection systems into its vehicles in late 1983. The names and number of computers it has employed over the years have frequently changed. Between the years 1983 and 1987, Chrysler's onboard computers were made up of two separate units: a logic module and a power module. The logic module was the smarter of the two, essentially serving as the brains of the system, since it contained a tiny microprocessor chip that received data from various sensors. The logic module was usually located on the right-hand passenger-side compartment, behind the kick panel. In contrast, the power module was typically located under the hood, near the battery, and it was a mere servant, taking its instructions from the logic module. The power module was responsible for switching actuators on and off, including the fuel injectors, ignition coil, and fuel pump. From 1987 to 1990, Chrysler introduced and used a single module engine controller model, or SMEC. This computer used two separate circuit boards housed inside a single plastic case. However, from 1989 to 1995, with improved technology the two circuit boards were combined into a single unit called a single board engine controller (SBEC). Even Chrysler's OBD-I dash warning light underwent name changes—initially called a power loss or power limited light, its name changed when the SMEC system came into being. Thereafter, the light was called a check-engine light—the same name employed by General Motors.

Unfortunately, the number of tests that can be performed, and the corresponding information that can be

This Chrysler power module was used to activate the ignition coil and other solenoids. It received instructions from a companion logic module (not shown). The large opening at left is the location of the air intake for the engine. When the vehicle was in use, moving air passed over the electronics inside the power module and kept everything cooled off.

generated using only a power loss or check-engine light on Chrysler vehicles, is somewhat limited—basically these systems are only capable of reading trouble codes. The good news is that if the scanner being used to test an older Chrysler OBD-I system has OBD-I capabilities and is used in conjunction with a Chrysler-compatible cable adapter, numerous functional tests can be performed as well as codes and sensor data accessed.

To access codes via use of the power loss light or check-engine light, turn the ignition key to ON and then to OFF, and then to ON and OFF again, and then finally leave it in the ON position. This will put the system into diagnostic mode. After accessing diagnostic mode, the power loss light will momentarily go out, and then come back on for about two seconds. The computer is performing a bulb check, which tests the computer's ability to control the warning light. There will be a brief pause, and then the light will begin pulsing out the diagnostic trouble codes, which will end with Code 55—indicating the end of the sequence. Later-model Chrysler vehicles equipped with an SMEC computer will pulse code 88, which will start the trouble code sequence; similar to the earlier bulb test, the code sequence will end with code 55. Any codes displayed between beginning code 88 and ending code 55 are trouble codes. All codes can subsequently be erased by disconnecting the battery for at least 10 seconds.

Once the computer has finished outputting all of the trouble codes, a switch test can be performed. Switched outputs include stepping on the brake pedal, turning the air conditioning switch on and off, or moving the transmission gear selector on a pre-1990 automatic transmission-equipped vehicle. Operating any switched output will cause the power loss light to either turn on or off. On vehicles equipped with a combination power module/logic module type of electronic computer system, rotating the front wheels—while at the same time counting the number of times the power loss light goes on or off—will test if the vehicle speed sensor is functioning. One wheel revolution should equal the equivalent of eight on/off cycles.

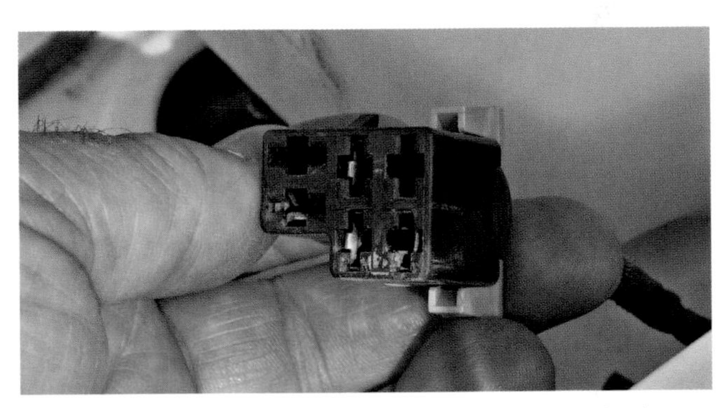

Pictured is Chrysler's OBD-I data link connector. Located under the hood in all Chrysler vehicles, it was oftentimes difficult to find. This connector was used from the early 1980s until it was eventually replaced in 1996 by the newer OBD-II DLC. *Courtesy Elwoods Auto Exchange, Hagerstown, Maryland*

With the use of a scan tool, an actuator test mode (ATM) can also be accessed on OBD-I-equipped Chrysler vehicles. In ATM, various actuators can be operated via inputs from the scan tool. For example, the ignition coil can be made to produce a spark, the fuel injectors can be cycled off and on, or the idle speed controllers can be made to operate, as can a whole host of other solenoids and relays. On later-model vehicles, even the gauges on the instrument panel can be operated with a scanner tool, along with many other components.

The End of OBD-I Diagnostics and the Future of OBD-II

By 1988, some version of the first generation of OBD-I computer systems was present in all cars and light trucks sold in the United States. However, as seen, there was no standardization between automakers with regards to OBD-I form, function, or technology. In fact, about the only thing the early OBD-I computers had in common with each other was that they all had some type of dash indicator warning light that was supposed to illuminate whenever the computer system detected a problem. The first generation of OBD-I computers was designed, in part, to let drivers know whenever a malfunction existed that affected the emissions coming from the exhaust tailpipe.

However, unlike the first generation of OBD-I computers, the second generation of OBD-II onboard computers now serve primarily as onboard emissions testers. Technology has improved to the point where the reliability of modern OBD-II computers assures functional monitoring of emissions systems every time the engine is started. These systems are so effective that many states have completely eliminated the old drive-thru, "sniff-the-tail-pipe" emission tests of yesteryear on those vehicle models manufactured in 1996 or later. Given modern OBD-II systems' abilities to detect and store information for subsequent retrieval, myriad problems ranging from a loose gas cap to a misfiring spark plug or disconnected vacuum hose to a bad computer sensor are simply and easily communicated to each state's emissions' testing program facilitators simply by the act of plugging in a universal OBD-II diagnostic link connector and watching the data stream pour forth. In addition to easier emissions testing, the second generation of OBD-II onboard computers provides a much-improved diagnostic tool for both professionals and home technicians alike, by enhancing their abilities to diagnose most engine performance-related problems.

Compared to professional scan tools of the past, not to mention most of the scan tools available today, this AutoXray CodeScout 700 is a bargain at around $90. It works on all vehicles from 1996 and later, and can read OBD-II and CAN diagnostic trouble codes, as well as erase codes, perform OBD-II monitor tests, and indicate if a vehicle will pass or fail an emissions test. *Courtesy AutoXray*

CHAPTER 2
OBD-II—THE ONBOARD EMISSIONS MONITOR TRANSITIONS

Transitions: From OBD-I to OBD-II

In 1988, the California Clean Air Act was signed into law by the governor of the state. This new law set forth all the various rules and regulations for statewide management of air quality for the next 20 years. As part of these new regulations, the specifications for standardization of second-generation onboard diagnostics (OBD-II) were implemented. Liking what it saw in California, the Congress subsequently amended its federal Clean Air Act in 1990, and in so doing, included California's provisions mandating standardization of OBD-II. The target date for uniform federal OBD-II implementation was 1996; the law specified that by this date all manufacturers had to achieve 100 percent OBD-II compliance for all vehicles sold in the United States. It's important to remember that OBD-II, as originally conceived, was not an engine management system, but rather, was simply a set of emissions-related rules and regulations that automakers were required to follow in the manufacture of all U. S.-bound passenger cars and light trucks.

The standardization of OBD-II technology came on the heels of prior efforts to control the quantity and quality of automobile emissions via the first generation of OBD-I computers. While the installation of OBD-I computers on some manufacturers' vehicles was certainly a step in the right direction, the consumer and governmental pressure placed on automobile manufacturers to limit auto emissions was not sufficient to force voluntary compliance within set guidelines, or at all, for that matter. Thus, automakers' efforts to curb emissions were not a big success, especially when it came to the practical reality of independent repair shops attempting to repair vehicles with malfunctioning emissions systems.

Because California's OBD-I state regulations were loose, to say the least, and in view of the fact that all other states did not have any regulations whatsoever that attempted to control or limit automobile emissions, manufacturers only did the bare minimum (more or less) required by the California Air Resources Board (CARB) and the EPA, and no more. With no uniformity between automakers, manufacturers devised their own technology and procedures for utilization of OBD-I systems to monitor emissions and to warn drivers of emission-related problems with their vehicles. Independent repair shops were faced with the prospect of having to purchase expensive diagnostic tools (scanners) and related computer software that only worked on a few models of vehicles they might be required to repair. Compounding the harsh economic reality facing independent repair shops who were trying to stay a step ahead of increasing OBD-I usage was the fact that different original equipment manufacturers (OEMs) required separate software cartridges as well as numerous specialized cables and adapters that had to be used in order to read a specific vehicle's OBD-I codes and data. The absence of readily available information, in the form of OEM repair manuals, parts' specifications, or manufacturers' recommended tests and procedures, only made matters worse.

Consequently, the $9 billion U.S. automotive aftermarket industry could see the handwriting on the wall if automakers were allowed to continue this trend to have it their own way. Obviously, there was greater economic incentive for automakers to keep everything on a proprietary basis, rather than to push the trend toward uniform standardization of onboard diagnostic systems in an effort to establish minimum emissions' controls standards. In this way, OEMs could control access to repair information and limit onboard vehicle communications (scanner interfaces) for their exclusive use, thereby essentially forcing automobile owners to take their vehicle(s) to a new car dealer for any and all emissions-related repairs. In addition, it was not inconceivable that auto manufacturers would have strong incentive to place individualized computer chip technology into all vehicle electronic components, thereby effectively cutting the aftermarket industry out of the electronic parts replacement marketplace, as well as severely limiting purchasing decisions for consumers.

Imagine purchasing a mass airflow sensor from a local auto parts store and installing it into a vehicle, only to learn after the fact that the vehicle's unique onboard computer searched in vain for an ID chip from Ford, GM, Chrysler, Toyota, Honda, or another auto manufacturer which it could not find. Consequently, the new airflow sensor would not be recognized by the OBD-I system as a suitable original equipment replacement part, and as a result, the OBD-I computer would turn on the check-engine light, or worse, prevent the car from starting altogether. While this may sound like a science fiction nightmare, the reality was that the technology that existed in the mid-1990s was specifically designed so that this repair/replace scenario, and others similar to it, would routinely occur, leaving the consumer confused, frustrated, and trapped, and at the mercy of auto manufacturers' increasingly secretive and ingenious design technology.

Fortunately, not all OEMs were as shortsighted as depicted in the previous example. In the early 1990s, Nissan and a few other manufacturers performed valuable marketing surveys on their existing customers and learned that if a customer's Nissan vehicle was easily repaired by a dealership, or more importantly, by an independent repair shop, the customer was more inclined to be satisfied with the purchase and to purchase another Nissan vehicle and/or recommend such a purchase to friends or family. Not surprisingly, vehicle ease-of-repair was a top factor in consumers' decision-making processes as to whether or not to purchase another vehicle of the same brand. These marketing surveys also forced auto manufacturers to face the flip side of the coin. If a particular vehicle proved to be difficult or costly to diagnose and/or repair, especially by the aftermarket repair industry, the customer was inclined to blame the manufacturer for the problem, and not the independent repair shop.

As a result of these marketing surveys, Nissan rose to the challenge and ensured Nissan training materials and Nissan-specific technical information were available to independent repair facilities. In the process, Nissan also provided the aftermarket industry with access to materials that were previously solely proprietary. Another more obvious factor that kept automakers from precluding the aftermarket repair industry from technical information about developing automotive computer technology was that the sheer number of cars and trucks requiring maintenance and/or repair was expanding. The economic good news for everyone involved was, quite simply, that if independent repair shops were all driven out of business and forced to close up shop, automotive dealerships would have gotten so overwhelmed with business that a minimum six-month waiting period for consumers to have vehicles repaired would have ensued. Ultimately, this would have led to extremely angry customers, who would naturally focus their anger on automotive dealerships and manufacturers.

Independent auto repair shops would have had an increasingly difficult time repairing customer vehicles if OEMs had not been "persuaded" by the government to share and standardize diagnostic and repair information. However, even today new car dealerships still have a slight advantage over independent repair shops, since proprietary factory codes and data are still far more comprehensive than what is available through the aftermarket industry. *Courtesy Keplinger's Automotive Center of Hagerstown, Maryland*

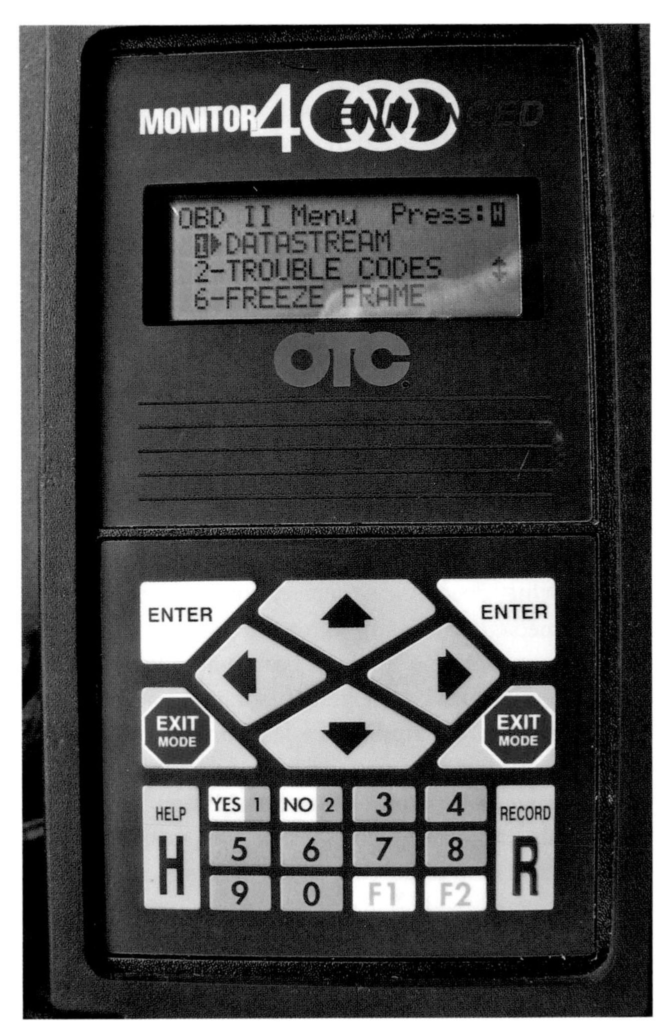

Scan tools like this OTC Monitor 4000E made it possible for independent auto repair facilities to compete with new car dealerships for vehicle owners' repair business. As a beneficial result, vehicle owners gained the option of having choices as to which facility to use to get a vehicle repaired. In addition to reading trouble codes and data streams, professional scan tools such as this also have tune-up specifications and troubleshooting tips stored within the scan tool. *Courtesy OTC Tools*

Fortunately for everyone, this potential nightmare never materialized. In the final OBD-II mandates enacted by Congress in 1997, auto manufacturers were required to provide the aftermarket industry with access to essential emissions control-related technical information and materials. A passage in the EPA Federal Register reads: "Manufacturers are required to make available to the aftermarket any and all information needed to make use of the OBD system, and such other information, including instructions for making emission-related repairs, excluding trade secrets." The legislative intent behind the phrase "any and

New car dealerships still have greater detailed diagnostic information available to them than is available to independent auto repair facilities, as proprietary codes and data streams can still only be read by OEM-approved equipment. Nevertheless, the ability to diagnose difficult emissions and engine performance problems has always come down to the skill and capabilities of the technician who is performing the work, and is not solely a function of the type of equipment used. *Courtesy Younger Toyota of Hagerstown, Maryland*

Today, standardization of automotive onboard computer diagnostic systems makes it possible for anyone with an OBD-II-compliant code reader or scan tool to take a peek inside a vehicle's power control module (PCM). Fortunately, it no longer matters if the vehicle being worked on is a General Motors, Ford, Jeep, Mazda, or even a Mercedes Benz model, as they all have the same standard diagnostic connector link (DCL) into which an OBD-II-compliant scanner can connect to. *Courtesy Mercedes Benz of Hagerstown, Maryland*

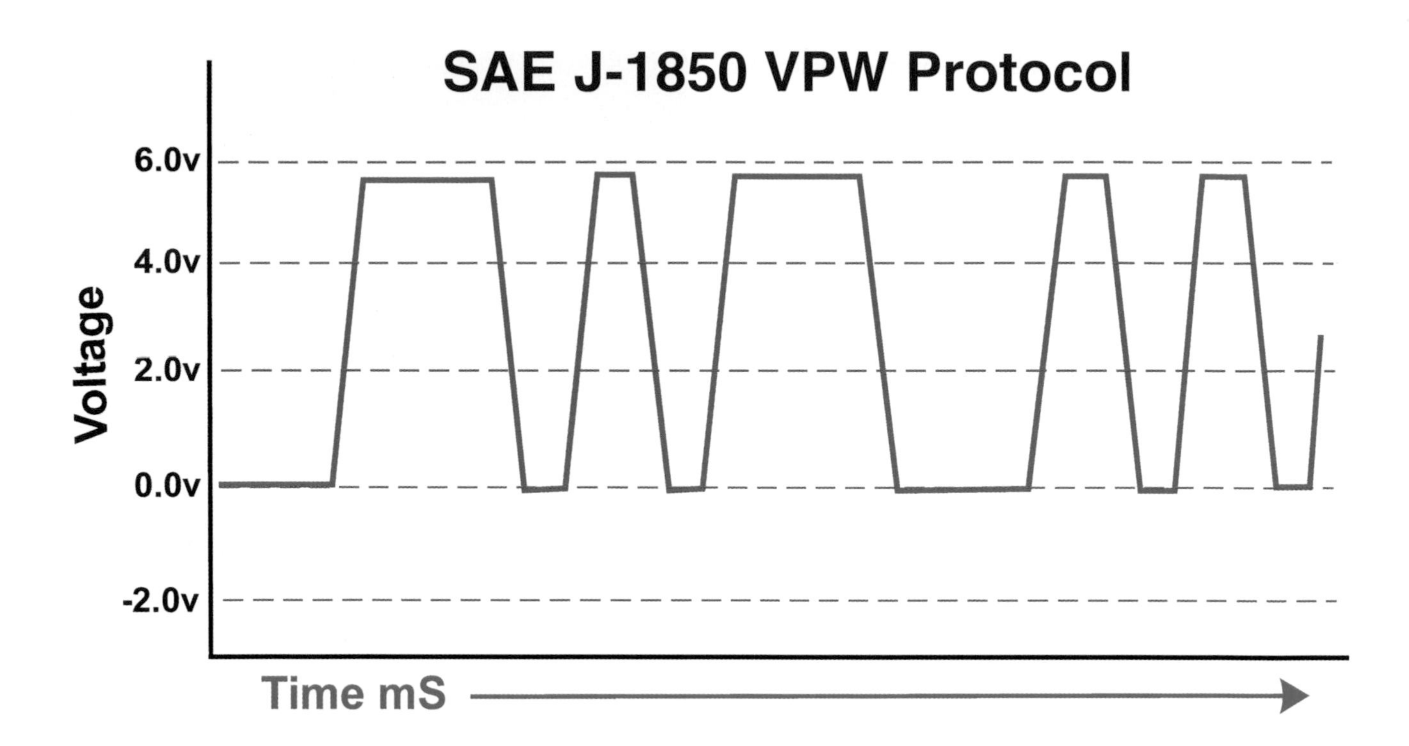

Fig 2-1—This is what communication between a powertrain control module (PCM) and a scan tool looks like on an oscilloscope. The waveform is a series of long and short pulses. This signal, also known as a bidirectional variable pulse width (VPW), can transmit data in two directions; however, because of its slow data rate and susceptibility to electrical interference, this protocol will eventually be replaced by a controller area network, or CAN, protocol.

exclusive means of digital communication between a scanner and OBD-II vehicle, communication methods had to be standardized. Standardization enabled scan tool manufacturers to design scanners compatible with any OBD-II vehicle. Scan tool communication is accomplished by a communication protocol that is governed by a set of rules and procedures that regulate data transmission between computers, as well as between test equipment (scan tools) and computers. These days, there are five communication protocols in use on OBD-II vehicles: (1) ISO-9141-2; (2) SAE J-1850 PWM; (3) SAE J-1850 VPM; (4) Keyword 2000; and (5) CAN, or controller area network. Fortunately, most scan tools automatically determine which protocol is in use, and are then able to immediately display codes and/or a data stream from a vehicle's powertrain control module (PCM). Following is a brief description of each protocol:

ISO-9141—The International Organization for Standardization 9141 protocol was used by Chrysler on all of their vehicles sold in the United States until 1998. It is still in use today on many cars. Most European and almost all Asian imports use this protocol. The speed at which data can be sent is 10.4 kilobits per second—or slow.

SAE J-1850 PWM—The Society of Automotive Engineers J-1850 Pulse Width Modulation (PWM) protocol is commonly used by Ford Motor Company on all of its domestic vehicles, as well as by Jaguar and Mazda. The speed at which data can be sent is 41.6 Kilobits per second—four times faster than the ISO protocol.

SAE J-1850 VWM—The Society of Automotive Engineers J-1850 Variable Pulse Width Modulation (VPM) protocol has been used by some Chrysler vehicles and most General Motors vehicles since 1998. Data transfer rates are 10.4 Kilobits per second (same as ISO).

Keyword 2000—The Keyword 2000 (ISO 14230) protocol is used mostly on vehicles manufactured and sold in Europe in the last year or two. Data transfer rates are 10.7 kilobits per second.

CAN—The Controller Area Network protocol was developed by Bosch and is a recognized international standard (ISO-11898) that is used on all 2004 and subsequent model years of Ford, Mazda, Mercedes, and Toyota vehicles. The CAN protocol will be mandatory on all vehicles by 2008. CAN is the hot rod of OBD-II protocols because its data transfer rate is one megabit per second—over 100 times faster than most other protocols.

CONTROLLER AREA NETWORK—CAN

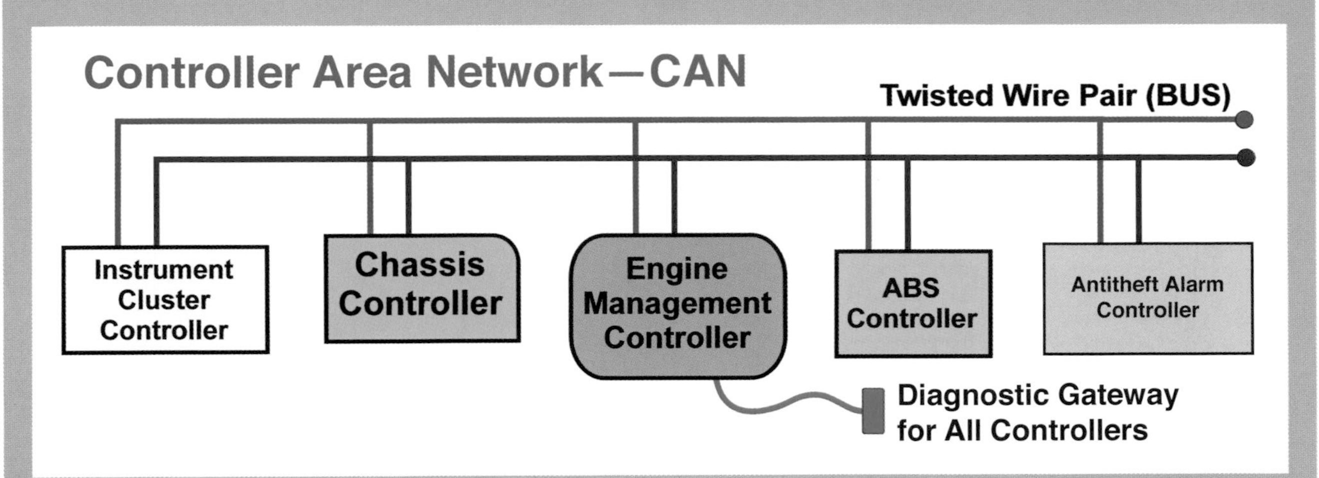

Fig 2-2—CAN will eventually change the way vehicle electronics are utilized. Smaller, lighter wiring harnesses and more reliable electrical connections will be used in contrast to the types of harnesses and connections typically found on current automotive electrical systems. Because CAN systems encompass a vehicle's entire electrical system, and can manage it with a single onboard computer, a higher level of diagnostic capabilities is possible. Both professional and do-it-yourself technicians will have to learn to adapt to this new technology.

A controller area network, or CAN, consists of multiple system components that use microcontrollers to communicate with each other. These communication systems are similar to integrated cable networks found in business offices, where desktop computers, file servers, printers, and digital phone systems all connect to and communicate with each other. However, CAN predates many office systems, as they were found as early as 1992 on some Mercedes Benz models, and 1993 on BMW 740i/iLs. Starting in 2003, CAN was used on some OBD-II vehicles to communicate with scan tools, and in 2004, Ford, Mazda, Mercedes, and Toyota equipped all of their OBD-II vehicles exclusively with CAN for all vehicle-to-scan-tool communications.

Developed by Intel, Bosch, and a few others, CAN communication technology has been around for about 15 years. The concept of how CAN works is simple: Computers, their sensors, and all power-consuming components on a vehicle are connected to each other via a single wire; this single wire is really two wires twisted together and is called a twisted pair. The twisted pair of wires a CAN system uses to communicate within the computer system is called a BUS, which is, quite simply, nothing more than a glorified communication network. The BUS allows all electronic information to be available at all times within the network of components, since digital messages are sent out by each computer or controller and received by all computers connected to the network.

For example, assume a driver wants to turn on a vehicle's high beams. A conventional automotive electrical system would operate as follows: The driver would activate the turn signal lever to switch on the high beams. This action would electrically trigger a relay that would, in turn, send power to the left and right high-beam headlights. Performing this same function on a CAN system actually happens quite differently, and with much more sophistication. When the driver switches on the high-beams, the chassis control computer receives the switched input and then triggers an internal power transistor (instead of a relay) that sends 12 volts to the high beams. But that's not all. In the event one of the high-beam bulbs burned out, the chassis controller would measure the current flowing to both lights (in this case, the current to the high beams would be half of its normal rate), and would then send out a message that one of the high beams was not working via a unique identifier code along the BUS wire. All controllers or computers in the system would get the message and check to see if the message ID code applied to them (and the systems they specifically control). If the ID code didn't apply to a particular computer or controller, the message would be ignored. However, when an instrument cluster controller receives the coded message, it recognizes the unique coded message as one that pertains to a system within its control, so it turns on a

continued on page 36

This is a PCM from a late-model Toyota. It uses the CAN protocol to communicate with a code reader or scan tool. CAN systems are able to send and receive information much faster than earlier OBD-II protocol systems. Most newer scan tools and code readers available are able to use the CAN protocol. *Courtesy Younger Toyota*

continued from page 35

warning light on the instrument panel informing the driver that there is a lighting system malfunction. In addition, an engine management controller also recognizes the message ID (one of the high beams is fried), and it also stores a trouble code in its memory for later retrieval by a technician using a scan tool connected to a diagnostic gateway.

Furthermore, significantly greater levels of diagnostics are available with CAN communication systems. For example, if a high-beam circuit had a wire that was shorted to ground, a CAN system would quickly measure the current going to the high-beam circuit, determine it is too high, and immediately shut off the power transistor that supplies power to the high beams before any damage could occur, thus preventing the chassis controller from moving into meltdown mode. In such a case, a different diagnostic message would be sent out across the network, since the code stored in the engine management computer might read "Excessive Current in High Beam Circuit." As a direct result of this superior level of circuit protection, no fuses have to be used in the entire automobile. In addition, controllers serve to switch power on and off to various components via transistors, instead of mechanical relays. Thus, the only relay used on an entire vehicle is a starter solenoid.

Another example of how a CAN-BUS system integrates functions from multiple controllers can be seen in an alarm

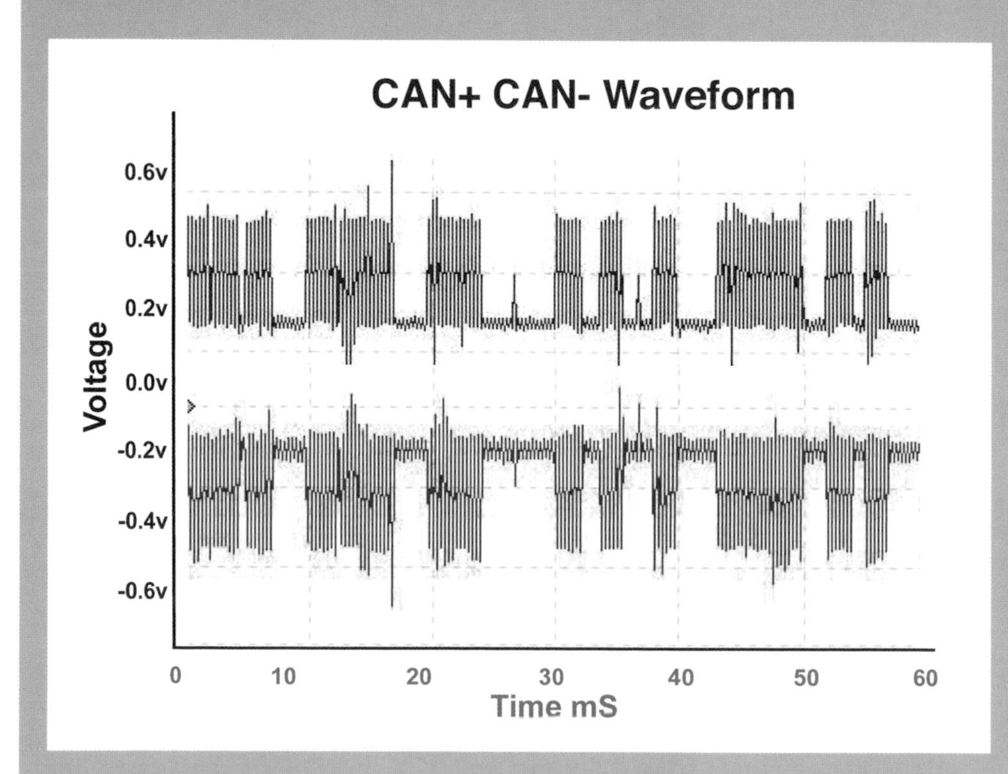

Fig 2-3—CAN uses two wires to communicate with a scan tool—CAN+ (top waveform) and CAN- (bottom waveform). Both wires use a bidirectional signal, which means messages can be sent in both directions. The CAN signal is like Morse code, in that the data is a series of long and short pulses. The two signals are an exact inversion of each other. This is called a differential BUS and it is resistant to electrical noise from outside and inside the vehicle.

system. If an alarm is set, and a would-be thief disturbs the car, several messages are sent out by the alarm controller: The first message instructs the chassis controller to turn on the hazard lights and sound the horn. A second message is sent to the engine management controller, which prevents operation of the electronic fuel pump. Once these messages are sent, no amount of hot wiring or ignition lock drilling will allow the engine to start or run.

Messages sent by controllers on some CAN systems are eight bytes long and travel at a speed of 500 k/bps (kilobytes per second) through the CAN-BUS system. This works out to roughly 4,500 messages per second. In fact, a CAN system can transmit data at a rate of one megabyte per second, or 30 pages of information each second. If two messages are sent at the same time, they are prioritized; one will be sent immediately and the other will be stored in the memory of the sending controller until BUS traffic allows it to be sent. Using a voltmeter to monitor CAN messages does not work, as it will only show 2.5 volts on the BUS wires when a message is present. Thus, a digital lab scope is the only way to actually see a message being sent.

While these systems are capable of performing self-diagnosis on many electrical-related problems, at least to some extent, they must still utilize wiring harnesses (though smaller) and connectors, all of which frequently have the same common electrical problems all technicians have come to know and love. Consequently, while the diagnostic capabilities a CAN system can offer will assist knowledgeable technicians with solving electrical problems, they will not actually replace repair technicians any time in the near future.

Another significant advantage of CAN systems over conventional automotive wiring is that fewer wires are used. For instance, the single wire BUS used on a CAN system utilizes shorter wires, which makes for a lighter wiring harness with more reliability. Additional benefits include less interference with low-voltage electronic fuel injection sensor signals, enhanced diagnostic functions for the entire electrical system, and available software updates whenever new electrical components are subsequently added. By 2008, all vehicles will be required to use some form of CAN in order to ensure standardized communication with scan tools and other diagnostic equipment. Eventually, entire electrical systems on cars and trucks will use CAN— it's only a matter of time.

The standard 16-pin OBD-II DLC is supposed to be located within 16 inches of the steering wheel. Unfortunately, that's not always the case. Some automakers do a good job of hiding them behind covers, under center consoles, and inside glove boxes. Fortunately, once the DLC is located, any OBD-II-capable scan tool or code reader can plug into it and start reading its data. If you're having a hard time finding the DLC in your vehicle, see the Appendix for a list of hard-to-find DLC locations.

OBD-II Data Link Connector—DLC

The location of the diagnostic link connector (DLC) on earlier OBD-I vehicles was often a mystery. Many manufacturers located the connector in roughly the same place on each model, but not always. Some manufacturers were truly innovative in their placement of the DLC—in fact, some of the more well-hidden are found (hopefully) on the 1994 Honda Prelude (under the passenger seat), the 1994 VW Passat (under the center console), and the 1993 Range Rover (under the passenger seat, tangled up with all the power seat wiring and the PCM).

With newer OBD-II systems, playing hide-and-seek with the DLC is not as much fun, or as difficult, as it used to be. Automakers must follow rules, specifically, standard SAE J-1962, regarding placement of a DLC. Every DLC must be in a centralized location within a vehicle's passenger compartment, and it must be within 16 inches (400 millimeters) of the steering wheel. Some manufacturers' interpretation of this standard is more loosely drawn than others, and as a result, some DLCs still remain well hidden. See the Appendix for a list of nonstandard locations for DLCs on various models.

Fortunately, once the DLC is located, they all have a uniform appearance and function, and all will connect to any OBD-II-compatible scan tool or code reader. Figure 2-4 (page 38) illustrates what the 16 pins in a DLC are configured for. Pins 1, 3, 8, 9, 11, 12, and 13 are not assigned to

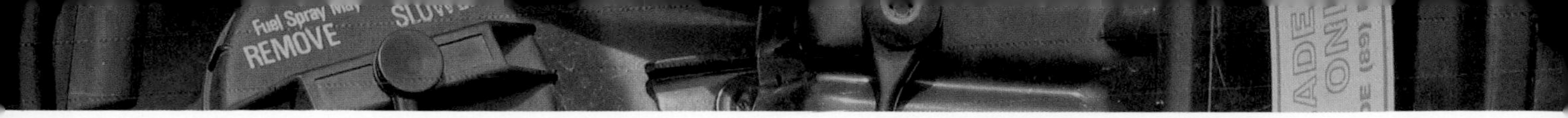

OBD-II Diagnostic Link Connector

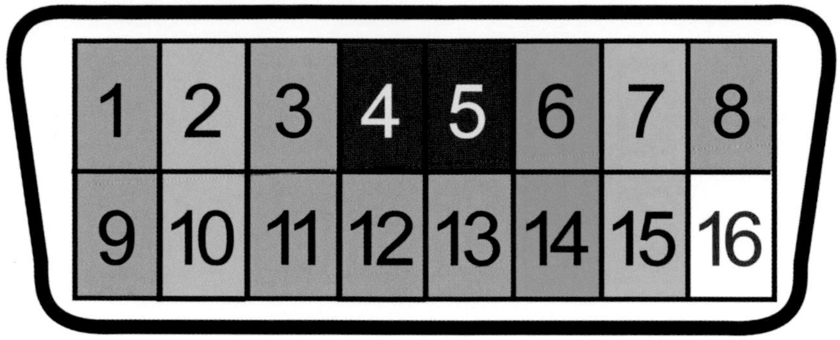

1	2	3	4	5	6	7	8
9	10	11	12	13	14	15	16

OEM Discretionary Use (unassigned)

Pin 2 BUS(+) Pin 10 BUS(-) (SAE J-1850)

Pin 4 Chassis Ground Pin 5 Signal Ground

Pin 6 CAN High Pin 14 CAN Low (SAE J-2284)

Pin 7 "K" Line Pin 15 "L" Line (ISO-9141)

Pin 16 Vehicle Battery Power

Fig 2-4—Almost all scan tools will automatically configure themselves to read whatever protocol a vehicle is using to send data to a scanner or code reader. Starting in 2003, some vehicles began using CAN systems as a method of communication. If a scan tool or code reader is not capable of reading CAN protocol, but you still want to determine if a vehicle actually uses CAN, a close look at a DLC will help. A DLC has 16 cavities that hold metal pins that serve as electrical connectors for a scan tool or code reader. Not all 16 cavities have pins present in them on all vehicles. Visually it can be determined which communication protocol a vehicle is using simply by looking at the pins inside the DLC.

If the pins that correspond with a specific communication protocol are populated (instead of blank—no pin), the vehicle uses that protocol. For example, if pins 5, 6, 14, and 16 are populated (the specific pins assigned to CAN communication protocol systems), the vehicle uses CAN protocol.

OBD-II functions; rather, these pins are used exclusively by manufacturers for their proprietary information or for enhanced diagnostics. Manufacturers will typically use some combination of these unassigned pins to communicate with dealership service department level scanners and other equipment that uses OEM software.

The remaining specifically assigned OBD-II pins can be used and accessed by everyone as follows:

Most European and almost all Asian vehicles use pins 4, 5, 7, 15, and 16 (ISO-9141) for their communication protocols;

Ford, Mazda, and Jaguar use pins 2, 4, 5, 10, and 16 (SAE J-1850 PWM) for their respective communication protocols;

Chrysler and most General Motors vehicles use pins 2, 4, 5, and 16 (SAE J-1850 VPM) for the same function; and

As of 2003, a limited number of manufacturers, including, Ford, GM, Mazda, and Saab, use pins 4, 5, 7, and 15 for their respective communication protocol. The trend toward using specifically assigned pins will continue until

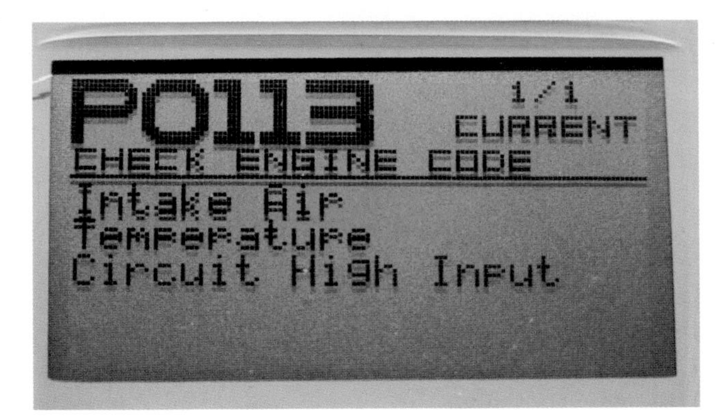

Whenever a scan tool or code reader is first purchased, it's a good idea to practice using it; this is easily done by making a PCM set a diagnostic trouble code (DTC). The first step is to disconnect an easy-to-reach sensor, then take the car for a short test drive. In this example, the air intake temperature sensor was disconnected. By connecting a scanner or code reader to the PCM, a DTC can be read. Voila! A P0113 DTC is displayed on the scan tool, along with the definition of the code: "Intake Air Temperature, Circuit High Input." After clearing the DTC and turning off the ignition key, the sensor should be plugged back in. This exercise will verify that both the scan tool and OBD-II system are operating as they should and provide you with some practice using the scan tool or code reader. *Courtesy AutoXray*

2008, when it will become mandatory for all manufacturers to use specifically assigned CAN pins.

A DLC has its own power source and grounds. Pins 4 and 5 are grounds, while pin 16 is battery power. These specific pins provide the scanner with both power and ground, so they don't require an external power source. Many scan tools use batteries so they can be used when not connected to a vehicle in order to store previous recorded data.

Diagnostic Trouble Codes—DTCs

OBD-II processes a staggering number of diagnostic trouble codes. There are over 4,000 generic diagnostic trouble codes (DTCs) just for the powertrain alone. This number doesn't include the thousands of codes for body control functions, lights, power windows, climate control, memory seats, mirrors, sound systems, antitheft alarms, or codes for the chassis control. Why are there so many? The high level of diagnostic detail requires such a voluminous number of powertrain DTCs.

For example, an onboard computer could use or set up to five DTCs for just one sensor, depending upon what's wrong with that circuit. For example, a manifold absolute pressure (MAP) sensor has five diagnostic trouble codes to cover a range of malfunctions, including:

- P0105—Manifold Absolute Pressure Circuit Malfunction
- P0106—Manifold Absolute Pressure Range/Performance Problem
- P0107—Manifold Absolute Pressure Low Input
- P0108—Manifold Absolute Pressure High Input
- P0109—Manifold Absolute Pressure Circuit Intermittent

Not only can an OBD-II computer tell if a MAP sensor is malfunctioning, it can also tell if it's disconnected, electrically open or shorted, is receiving no engine vacuum, or its output signal logically disagrees with other sensor inputs. In general, OBD-I vehicles previously had only one code for a MAP sensor, not providing enough information to offer much of a diagnosis. In the early days of OBD-I diagnostics, a technician could simply memorize the majority of common trouble codes instead of having to look them all up in a service manual, especially at dealerships where only one brand of vehicle was being repaired. Obviously, that's not the case today. With thousands of codes available to technicians today, they must have access to good sources of technical information and code definitions. In addition to displaying codes, many scan tools, and even some code readers, can display trouble code definitions, as well as some repair tips and tune-up specifications. Definitions for DTCs are also available from scan tool manufacturers via the Internet. There is also an abbreviated list of generic OBD-II DTCs in the Appendix of this book.

All generic OBD-II diagnostic trouble codes use five alphanumeric characters for identification. On all OBD-II DTCs, the first character on the left (as shown in Figure 2-5 on page 40) identifies in which operating system within the vehicle the DTC is showing a malfunction for. There are four basic operational systems which have diagnostic trouble codes: P represents the powertrain system (engine and transmission); B represents body including air bag, instrument cluster, lighting, power seats, memory mirror); C stands for chassis (ABS brakes and traction control); and U represents network communication.

The second character from the left always indicates whether the DTC is a generic OBD-II code that everyone can access, or a specialized manufacturer's code that only technicians associated with that specific manufacturer can exclusively access. When the SAE came up with the original DTC list, many manufacturers complained that they wanted to use their own code numbering systems. Consequently, the SAE indulged the manufacturers to some extent by separating standard OBD-II generic codes (P0 codes) from manufacturer-specific P1 codes. All auto manufacturers have exclusive

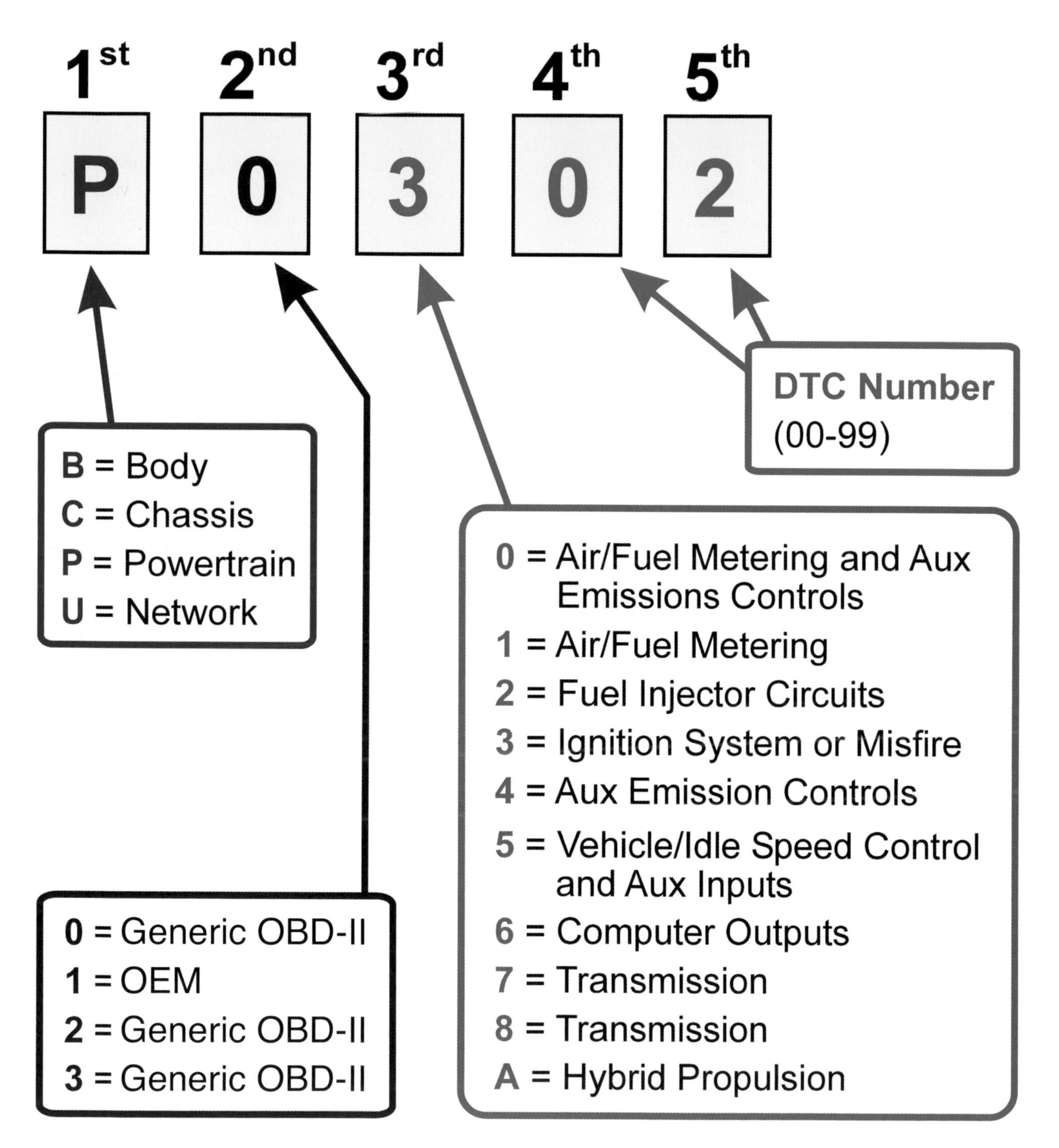

Fig 2-5—Standardization of DTCs make this P0302 DTC easy to read. The "P" indicates a powertrain code (indicating a malfunction exists somewhere in the powertrain system), "0" indicates a generic OBD-II code that everyone can access (as opposed to a manufacturer-specific code that only dealerships associated with that specific manufacturer could access), the "3" reveals specifically an ignition system misfire, and the last two characters (0 and 2) confirm cylinder number two is misfiring.

scan tools and other diagnostic equipment capable of reading their own proprietary codes and other proprietary data, which are above and beyond the generic OBD-II codes and data streams that typical aftermarket scanners and code readers can read. The additional proprietary information that manufacturers scan tools are capable of reading is known as enhanced diagnostics. DTCs with a second character of 0, 2, or 3 are generic OBD-II codes. Any DTC with a 1 for its second character is a manufacturer-specific original equipment manufacture (OEM) code; all the rest are generic OBD-II codes, which any scan tool can access. P1 fault codes are primarily proprietary for manufacturer's use only; however, certain sophisticated scan tools can interpret some of them.

The third character position in a DTC indicates which specific system is experiencing a malfunction. The malfunctioning system sets a system-specific trouble code. In any OBD-II DTC, this character really provides the most useful information to a technician since the general location or system within a vehicle that is experiencing a malfunction can be identified without even having to look up the DTC. The third character system code numbers are as follows:

- Faults relating to the air/fuel control systems
- Faults relating to the fuel system's fuel injectors
- Faults relating to the ignition system (including misfires)
- Faults relating to auxiliary emission controls, including EGR, AIR, CAT, and EVAP
- Faults relating to vehicle speed, idle control, or auxiliary inputs
- Faults relating to computer systems, PCM, or CAN
- Faults relating to the transmission or transaxle
- Faults relating to hybrid propulsion-powered vehicles

The fourth and fifth characters in a DTC are similar to older OBD-I codes, in that both digits are read together to ascertain more system-specific information about the actual component that is experiencing a malfunction. These characters display a specific code, which represents a malfunction within a system or a component within a system, which the PCM communicates whenever it detects a malfunction or fault.

Freeze-Frame Data

In addition to detecting and displaying DTCs, OBD-II systems also feature another diagnostic tool known as freez-frame data. Every time a DTC is set, the equivalent of a snapshot of overall generic emissions information

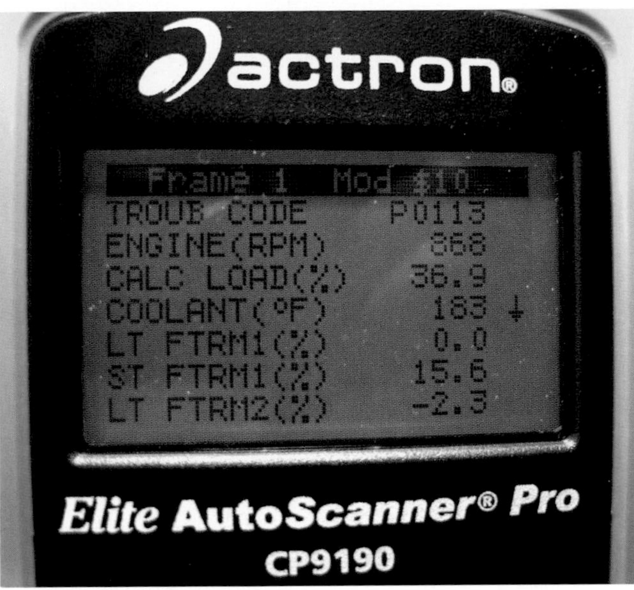

This Actron scan tool is displaying freeze frame data captured from when a PCM set DTC P0113. Freeze frame data is useful when diagnosing why a code was set in the first place, since specific operating conditions are registered and stored at the exact moment a malfunction occurred. Freeze frame data is erased on most vehicles whenever the DTC is erased or cleared, so it's always a good idea to write down all stored data, including freeze frame data, before erasing any trouble codes. *Courtesy Actron*

stemming from engine operating conditions and other data are taken and stored in a PCM's computer memory. Freeze-frame data should not be confused with a similar function known as a failure record. A failure record is another type of snapshot equivalent that is recorded at the time a specific malfunction last occurred, and then subsequently updated each time the same specific malfunction occurs again. Not all auto manufacturers use failure records in their OBD-II computer software, but they all definitely use freeze-frame data.

Whenever a DTC is set on most vehicles (though not all), freeze-frame data is registered and recorded for subsequent retrieval by a technician. Freeze-frame data typically includes the following information:

- DTC number
- Open or closed loop status
- Engine coolant temperature (ECT)
- Vehicle speed (VS)
- Intake air temperature (IAT)
- Engine rpm
- Misfire data (specific misfiring cylinders may be listed)
- Calculated engine load percentage

- Manifold airflow pressure (MAP)
- Fuel pressure (if controlled by the PCM)
- Up and down stream oxygen sensor (HO_2S) voltage (listed for each bank of cylinders)
- Short- and long-term fuel trim (STFT—LTFT) percentage
- Total number of DTCs stored in memory

Some automakers provide more freeze-frame data than many scan tools are capable of reading; consequently, not all scan tools can display all freeze-frame data information. However, to the extent available, freeze-frame data is valuable to a technician attempting to duplicate similar conditions during road testing to determine if an emissions problem really has been fixed. In addition, intermittent failures that are normally hard to reproduce can be reproduced more easily by using freeze-frame data. Furthermore, since certain failures in one emissions system can cause or contribute to problems in another system, freeze-frame data allows a technician to compare readings between various engine operating systems to determine if any of the information doesn't make sense or is inconsistent.

OBD-II Monitors

How does an OBD-II system determine when to set a particular DTC? It runs up to 12 different diagnostic tests, called monitors, which check all systems within its control for any malfunctions that potentially could affect emissions. Sometimes these monitors are referred to as inspection and maintenance (I/M) readiness tests or readiness flags. Monitors are divided into two groups: continuous monitors and noncontinuous monitors.

Continuous monitors are the more important of the two, as any failure or malfunction in any system associated with these monitors would have more serious consequences for emissions. Continuous monitors help protect a catalytic converter from damage, so it can functionally reduce emissions. Consequently, these monitors are highest priority for OBD-II systems. Excess fuel delivery or ignition misfires are a catalytic converter's worst nightmare, since either of these conditions will cause enough damage to make cat replacement necessary. Continuous monitors start running as soon as enabling criteria, or engine-operating parameters, are met. Enabling criteria are simply minimum engine operating conditions that must be met in order for a continuous monitor to run. For example, an engine must be at a minimum threshold operating temperature before a catalyst monitor can run.

Noncontinuous monitors are not as critical as continuous ones, as they don't have the same effect on emissions and don't run as often as continuous monitors; this is because the minimum enabling criteria for continuous monitors is much less extensive (or lower) than for noncontinuous monitors, and consequently, the lower minimum threshold for operation causes them to run most of the time the vehicle is operational. In contrast, noncontinuous monitors may run only occasionally, and not necessarily each time the vehicle is driven, again because they don't have the same significant effect on emissions, as do continuous monitors.

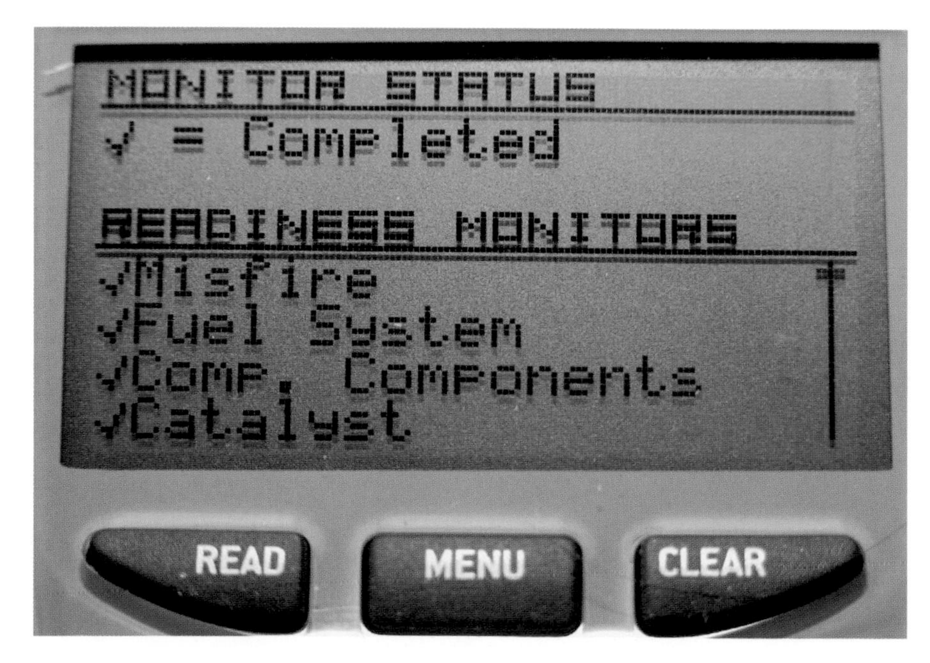

The status of OBD-II I/M readiness monitors is shown on this AutoXray CodeScout 2500 code reader. System monitors with a checkmark next to them are completed, which simply means a PCM has successfully conducted testing on these components used by that monitor. In the instance shown, the misfire, fuel system, comprehensive component, and catalyst monitors have all finished operational testing. The scroll bar on the right side of the display lets users scroll down to see the rest of the monitors and their status. *Courtesy AutoXray*

Of the 12 OBD-II system monitors, the first three are continuous monitors. All the rest are noncontinuous monitors. The last three noncontinuous monitors are newer in existence, and are therefore oftentimes not found on earlier OBD-II vehicles. The 12 OBD-II monitors are as follows:

- Ignition misfire — *Continuous*
- Fuel system — *Continuous*
- Comprehensive component monitor — *Continuous*
- Catalyst efficiency
- Heated catalyst
- EGR system
- Evaporative system
- Heated oxygen sensor
- Secondary air injection
- Air conditioning
- Positive crankcase ventilation
- Thermostat

The PCM has to run all of the monitors while at the same time managing the engine's fuel delivery and ignition systems. This software configuration is continuously running in the background, invisible to a technician, with regard to both OBD-II diagnostics and engine systems management. However, this organizational software does not assist with setting DTCs or other OBD-II functions. OBD-II monitors perform software calculations for their own operation thousands of times per second, so something must be in charge of overseeing all these tasks in order to keep things occurring in orderly fashion. Just as a conductor of a symphony orchestra keeps 100 musicians in sync with the score of the music, OBD-II systems must keep monitors coordinated and prevented from interfering with each other and with engine management systems. General Motors and Ford call this software, which is the equivalent of an OBD-II system conductor, a Diagnostic Executive, while Chrysler refers to it as a Task Manager. A Task Manager or Diagnostic Executive performs all of the following functions:

- Starts and runs the monitors
- Prioritizes the monitors
- Updates the readiness status for the monitors
- Prevents conflicts between any of the monitors
- Directs the monitors to run in the correct order
- Manages the operation of the monitors and keeps them from interfering with engine management
- Displays test results for all monitors on a scan tool
- Keeps track of the drive cycles that control DTCs
- Stores DTCs, freeze-frame data, and controls the status of the malfunction indicator lamp

A Task Manager places the monitors in one of four modes of operation: normal, pending, conflict, and suspend. Monitors only run when they are in normal mode; any other mode will cause the monitor(s) to postpone testing.

Pending Mode—A monitor is placed in pending mode whenever a signal from a sensor is discrepant from what the PCM is programmed to expect. The monitoring test will be postponed pending repair of the malfunction. For example, if an oxygen sensor's voltage levels were too high or too low, the Task Manager will not perform a catalyst efficiency monitor test, as this test requires proper functioning of front and rear O_2 sensors.

Conflict Mode—A monitor is placed in conflict mode whenever the simultaneous operation of two monitors could possibly conflict with each other. One monitor will be delayed until the other is finished testing. For example, if a PCM commands an EGR valve to open while performing an EGR system monitor, the comprehensive component monitor for the idle air controller (IAC) cannot also be run at the same time, since EGR operation would affect an IAC actuator's performance.

Suspend Mode—A monitor is placed in suspend mode whenever it is delayed to allow proper sequencing of tests in relation to other monitors and enabling criteria. For example, if a PCM needed to run an evaporative system monitor, but the fuel tank was almost empty, an EVAP monitor would be suspended or delayed from performance until the gas tank was restored to between 20 and 80 percent full.

Drive Cycles and Enabling Criteria

In order for a monitor to start its programmed test, certain minimum enabling criteria must be met. These enabling criteria, or minimum threshold operating conditions, are collectively called a drive cycle or trip, and amount to nothing more than simply starting the vehicle when it is cold and driving it until the engine reaches normal operating temperature, then driving at different speeds, and finally turning the ignition off. Some monitors require only one complete drive cycle before a test can start, while other monitors may require more than one drive cycle or trip. If the minimum requirements of a drive cycle are not met, a monitor cannot run its test(s), and therefore, are unable to make any determination as to whether or not a malfunction exists within components or subsystems tested by that monitor.

Each monitor has two readiness states: quite simply, either ready or not ready. When a monitor status is ready, the required drive cycle has been performed and the monitor has successfully completed running its tests. When monitor status is not ready, the drive cycle for that monitor has not been successfully completed, and therefore, the monitor cannot run the tests for its system. EPA guidelines specifically allow up to two monitors to be in a not-ready state during vehicle operation for model years 1996 through 2000, but only one monitor for 2001 and newer models. Some scan tools and code readers refer to "complete" and "incomplete" in place of "ready" or "not ready."

A generic OBD-II drive cycle exists that is supposed to run all monitors. Unfortunately, this generic drive cycle doesn't always work as it should, since many auto manufacturers require very specific combinations of engine temperature, speed, and load changes to be present in order to trigger a particular monitor. Quite frankly, nothing beats access to genuine factory service information when trying to convince an OBD-II system to run its monitors. The following represents a typical generic OBD-II drive cycle or trip:

- The fuel tank must be between one-quarter and three-quarters full to run the EVAP monitor.
- The ignition key must not be left on prior to cold starting the engine, or the heated oxygen sensor monitor may not run.
- The vehicle's engine has to be cold to start the drive cycle; cold is defined as coolant temperature below 122 degrees Fahrenheit, with coolant and air tem perature sensors reading within 11 degrees of each other.
- Once the engine starts, the automatic transmission is placed in drive and the engine allowed to idle for 2 1/2 minutes with air conditioner and rear defrost on.
- The A/C and rear defrost are turned off, and acceleration is gradually (moderately) increased to 55 miles per hour until a steady speed of 55 is held for five minutes.
- Deceleration (coasting down) to 20 miles per hour without braking or depressing the clutch on a manual transmission occurs next.
- Acceleration back to between 55 and 60 miles per hour at half throttle follows next, and this speed is held for five minutes.
- Finally, deceleration (coasting down) to 20 miles per hour without braking occurs. The vehicle is stopped via the brakes and the engine is allowed to idle for 10 seconds.

By performing this generic OBD-II drive cycle (or a manufacturer's specific drive cycle), all the monitors should be ready to run and complete their tests. If any malfunctions in any systems are found, the appropriate DTCs will be set and displayed once a scan tool is connected to a PCM. Some monitors may require two consecutive drive cycles or trips before they can actually be run. For more information on manufacturer-specific drive cycles, see Chapter 7.

Misfire Monitor

A misfire monitor continuously checks for engine cylinder misfires. Because of the potential severity of damage caused to a catalytic converter by ignition misfires and raw fuel subsequently entering the converter, this monitor is continuous and runs most of the time the vehicle is being driven. Additionally, this monitor doesn't require as much minimum threshold enabling criteria as other monitors; in other words, as soon as the engine is started on most vehicles, a misfire monitor starts running. While many misfires can be caused

Monitor	A	B	C	D	E	F
Ignition Misfire (types 1 & 3)	Continuous	1	2	1	3 Trips	80
Ignition Misfire (types 2- severe)	Continuous	1	1	1	3 Trips	80
Fuel System	Continuous	1	1 or 2	1	3 Trips	80
Comprehensive Component	Continuous	1	2	1	3 Trips	40
Catalyst Efficiency	Once per Trip	1	2	1	3 Trips	40
Heated Catalyst	Once per Trip	1	2	1	3 Trips	40
Exhaust Gas Recirculation (EGR)	Once per Trip	1	2	1	3 Trips	40
Evaporative System	Once per Trip	1	2	1	3 Trips	40
Heated Oxygen Sensor	Once per Trip	1	2	1	3 Trips	40
Secondary Air Injection	Once per Trip	1	2	1	3 Trips	40

A Times monitor runs per trip
B Number of trips required, with a fault prsent, to set a pending DTC
C Number of trips required, with a fault prsent, to command the MIL "On" and store a DTC
D Number of trips required, with a fault prsent, to erase a pending DTC
E Number of trips required, with a fault prsent, to turn off the MIL
F Number of warm-up cycles required to erase the DTC after the MIL is turned off

Fig 2-6—The chart shown lists information about each respective monitor in an OBD-II system. Continuous monitors are graphically shown in yellow, while noncontinuous monitors are listed below in orange. Column F represents the number of warm-up cycles it takes a PCM to erase all DTCs. In general, warm-up cycles are defined as occurring whenever engine temperature is below 122 degrees Fahrenheit on start-up and coolant and air temperatures are within 11 degrees Fahrenheit of each other during a cold start.

The General Motors crank sensor on the left and cam sensor on the right provide crankshaft position and cylinder-firing-sequence information to a vehicle's PCM. In addition to triggering the ignition system's firing of spark plugs, a crank sensor can also detect ignition misfires by monitoring the rotational speed of the crankshaft. If engine speed slows for even a fraction of a second, a code for an ignition misfire may be set, even including in which cylinder the misfire is occurring.

by an ignition system, other malfunctions can also cause them as well, including vacuum leaks, improper EGR valve operation, air-fuel ratio problems, incorrect ignition timing caused by a faulty cam belt, fuel injector malfunctions, and a sticking PCV valve. All of these potential malfunctions can set a P0300 DTC—which detects a "random/multiple cylinder misfire." If the misfire code displayed has a number other than 0 in either of the last two digit positions, such as a P0305, then a specific cylinder is misfiring, and is specified by the last number (in this case, cylinder 5).

Misfires are detected whenever a PCM measures engine speed fluctuations via either a cam sensor or crankshaft speed sensor. The rotational speed of the engine, caused by each cylinder's power stroke, is measured and then compared with the power stroke that precedes it and follows it—before and after the misfire occurs. If a misfire occurs in a cylinder, that cylinder will not push as hard on the crankshaft, and engine speed slows for a fraction of a second. A PCM will measure this pause in engine speed and set a misfire DTC.

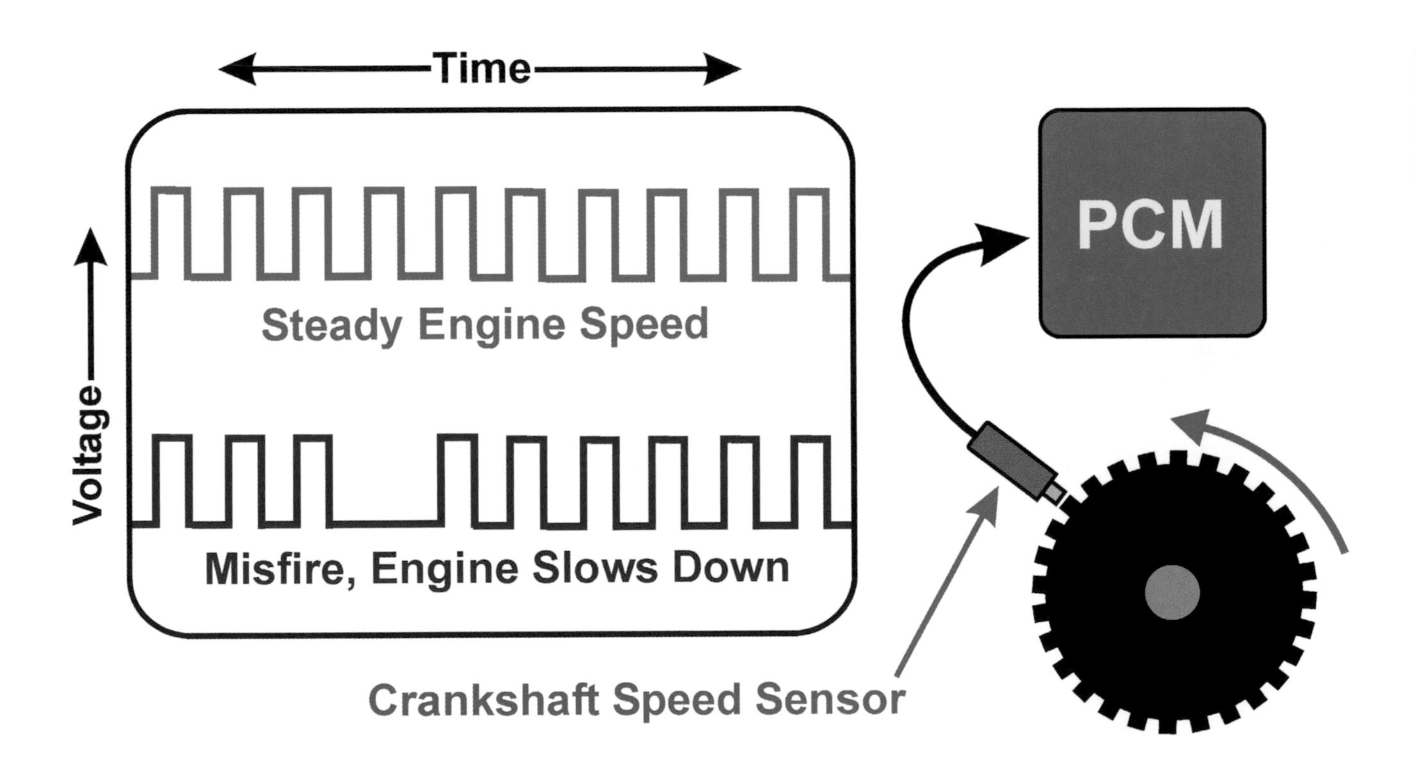

Fig 2-7—A PCM monitors engine speed via a cam or crankshaft sensor. The waveform in green (top of diagram) indicates a steady engine speed—thus, no misfire. If an ignition-related misfire or other type of misfire occurs, the engine's rotational speed slows for a fraction of a second because no power stroke from the misfiring cylinder is received (waveform in red at bottom of diagram). Not only can a PCM detect a misfire, it can also determine which specific cylinder is misfiring based on the crankshaft's position when the misfire occurs. In addition, sometimes a PCM will set a misfire code because of an improperly balanced tire and wheel, since this condition could change the speed of an engine in a manner similar to that of a true misfire.

There are three kinds of misfires. They are known as Type 1, 2, and 3. Type 1 and 3 misfires are "two-trip" monitor malfunctions. Two-trip monitor malfunctions occur whenever a misfire is detected on a first trip or drive cycle, and the PCM then records and stores the malfunction as a pending diagnostic trouble code, without turning on the MIL until the malfunction occurs again during a second trip or drive cycle. If the misfire fault happens a second time under similar driving conditions of engine speed, load, and temperature, the MIL will be turned on and a DTC set. In essence, the PCM wants to see the problem occurs more than once before letting a driver know there is a malfunction in a system which has been set and displayed as a DTC.

Type 2 misfires are the most serious as they cause damage to a catalytic converter. When a type 2 misfire is detected, the MIL comes on immediately; if the problem persists, the MIL will flash once each second, continuously. In addition to real misfires, false misfires can also turn on an MIL and set a DTC. False misfires can be caused by a worn or loose camshaft-timing belt or by driving a vehicle on a ridged road with a washboard surface. Both of these conditions can have an effect on crankshaft speed, and could cause a PCM to set a false trouble code or inappropriately turn on the malfunction indicator light.

Comprehensive Component Monitor (CCM)

The comprehensive component monitor, or CCM, checks input and output components for the presence of electrical shorts and opens, and monitors systems within its control for any out-of-range values. In addition, it also checks for functionality and rationality. Just as earlier OBD-I systems checked individual sensors for electrical values that were out of range, OBD-II systems have a similar function via

Comprehensive Component Monitoring

Input Components:

MAF Sensor	Vehicle Speed Sensor
Knock Sensor	Intake Air Temperature
MAP Sensor	Cam Position Sensor
Brake Sw	Coolant Temperature
Gear Selector Sw	Throttle Position Sensor
Cruise Control Sw	Crank Angle Sensor
Fuel Type Sensor	

Output Components:

Idle Speed Control	Warm-up Catalyst Bypass Valve
Heated Fuel Systems	Electronic Transmission Control
Glow Lamp (Diesels)	

Fig 2-8—Similar in some respects to older OBD-I diagnostics, a comprehensive component monitor (CCM) continuously watches various sensors and controls to make sure they're all doing what they're supposed to do. However, unlike less sophisticated OBD-I systems, the CCM on an OBD-II system also takes input and output logic into account as it evaluates component performance. If sensor readings don't make sense to a CCM, it will record a malfunction; if the malfunction is perceived as serious enough, or persistent, the CCM will turn on the service engine soon light, or MIL.

This coolant sensor is just one of many computer sensors monitored by a CCM. Some OBD-II systems are capable of diagnosing a stuck-open thermostat simply by monitoring the rate at which the coolant temperature rises after engine start-up. If coolant temperature does not reach a predetermined point after so many minutes of engine operation, a DTC could be set to indicate a problem with the engine's thermostat. *Courtesy Elwoods Auto Exchange*

This Ford throttle position sensor (TPS) provides the onboard computer with driver demand information. Specifically, the information relayed confirms how hard or rapidly a gas pedal is being pressed. The TPS output signal has to correlate with other sensors, such as a mass airflow (MAF) sensor and a manifold absolute pressure (MAP) sensor. For example, if the TPS sensor shows that the throttle is wide open, then the MAP or MAF sensors should indicate that the engine is operating under maximum load and producing maximum torque. If any of these sensors does not logically agree with the output of the others, the comprehensive component monitor may set a DTC to indicate a malfunction.

the operation of a continuous monitor. Fortunately, OBD-II systems have additional advantages over OBD-I systems; a CCM also monitors the length of time it takes various sensors to meet minimum enabling criteria after engine start-up, and monitors sensor inputs to ensure they correlate and make sense in their relationship with one another.

For example, if an engine coolant temperature (ECT) sensor and an intake air temperature (IAT) sensor both read 55 degrees Fahrenheit when the engine first starts up, a PCM would expect the ECT to increase at a fairly rapid rate as the engine warms up. If, after five minutes, the ECT only reads 70 degrees Fahrenheit, the PCM would set a

PCM Fuel Correction

RPM \ Engine Load %	10	20	30	40	50	60	70	80	90	100
5500	0	-1	-2	0	-5	-6	0	+2	+5	-7
5000	+6	+3	-8	-7	-3	+4	+3	-1	0	+1
4500	-6	-5	+4	+3	-2	0	+1	-4	+2	-2
4000	-4	-3	+1	+5	+2	-1	-2	-8	-5	+4
3500	+6	-2	-8	-7	+3	-4	+3	+6	+5	-3
3000	+2	-6	-3	+2	+3	-6	0	-5	-2	0
2500	0	-5	+1	**+3**	0	-2	-1	+2	+5	-3
2000	-8	-6	-2	+4	+5	-4	+6	-3	-8	-4
1500	+4	+5	-7	+3	-5	-8	+1	+3	+4	+6
1000	+3	+1	-1	-2	-4	0	+2	0	-7	-6

Fig 2-9—This chart illustrates how a PCM's adaptive learning strategy continuously adjusts to correct fuel delivery. Every single combination of engine rpm and engine load condition has a number assigned to it. (Engine load is the amount of power an engine has to make to run a vehicle.) As operating conditions change, the fuel correction number also changes, based upon an oxygen sensor's input to the PCM. In the example shown in the graph, the adaptively learned number "+3" means that at 2,500 rpm with a 40 percent engine load, a fuel correction factor of "+3" is required to maintain the correct air/fuel ratio for the catalytic converter. Adaptive learning allows the PCM to constantly adjust the air/fuel ratio to match any specific individual engine cylinder's mechanical wear, as well as dimensional tolerances found in the manufacturing process.

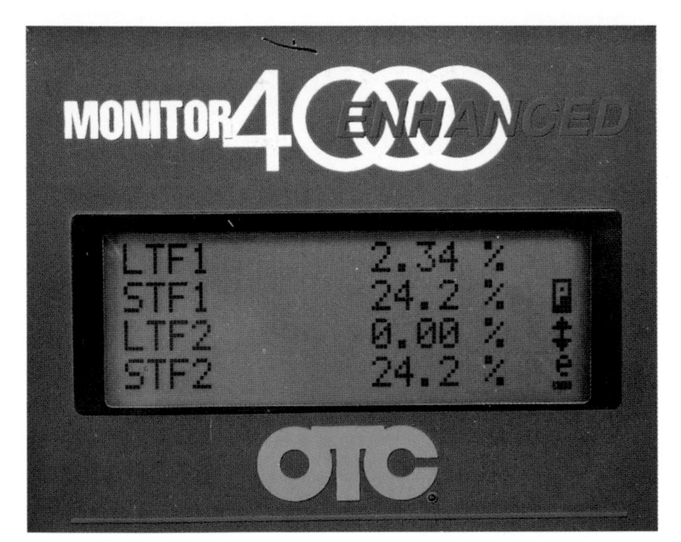

This scanner is displaying both long-term and short-term fuel trim for a V-8 engine. The OBD-II system uses two oxygen sensors (one for each side or bank of a V-8 engine) to monitor exhaust gas oxygen content. Based on the information received from the oxygen sensors, fuel trim calculations are processed. Fuel trim is expressed in percentages, with 0 percent indicating no fuel correction is required for a specific engine rpm and calculated load value. LTF1 stands for long-term fuel trim on bank 1 (one side of the V-8), and STF1 stands for short-term fuel trim on bank 1 (the same side). LTF2 and STF2 represent the long-term and short-term fuel trim percentages for the other bank of cylinders. *Courtesy OTC Tools*

P0116 DTC, which represents an "Engine Coolant Temperature Circuit Range/Performance" malfunction. However, if after two minutes of driving the engine temperature rises to significantly hotter than 250 degrees, an ECT reading would not be logical since the engine cannot possibly warm up that quickly. Another example that illustrates this point happens whenever a throttle position sensor (TPS) reads three-quarters open, but the engine rpm is only at 600 at idle. Thus, the TPS sensor would fail the CCM monitor's rationality test, because the TPS values for throttle opening don't actually match the low engine rpm.

Fuel System Monitor

A fuel system monitor uses the adaptive fuel corrections programs stored inside the PCM to monitor the fuel system and engine performance. This continuous monitor requires more minimum enabling criteria be met than a misfire monitor does. The PCM on OBD-II vehicles uses the strategy of adaptive learning to constantly adjust and correct fuel delivery to the engine. Two programs are used to accomplish fuel correction—short-term fuel trim (STFT) and long-term fuel trim (LTFT). Together, these programs keep the air-fuel ratio close to 14.7:1, the ideal ratio to enable a catalytic converter to operate at optimum efficiency. More information on air-fuel ratios and catalytic converter operation can be found in Chapter 3.

Short-term fuel trim (STFT) is constantly changing, based on an oxygen sensor's input to the PCM. If the exhaust gas oxygen content indicates a lean air-fuel ratio condition, the STFT is adjusted to add more fuel to the engine. STFT uses a software program that stores information in different cells that correspond with each specific different engine speed and engine load combination possible (see Figure 2-9). (Engine load simply represents how hard the engine has to work, or how much power it must produce to accelerate or maintain vehicle speed.) As driving conditions change, STFT is adapted to correct fuel delivery. For example, if engine speed is 2,500 rpm, with an engine load of 40 percent, the STFT could be +3 percent, indicating that at that speed and engine load an additional 3 percent more fuel needs to be added to keep a catalytic converter functioning properly. If, under the same conditions, an O_2 sensor indicates a rich air-fuel mixture, the STFT would reduce its correction factor by minus 2 percent. STFT continues producing continuous adjustments or corrections in the air-fuel ratio as long as the engine is running. STFT is expressed on a scan tool as a percentage of (+/-) 10 percent.

The LTFT only adjusts or changes its own fuel correction values whenever the STFT adjusts air-fuel ratio levels past certain predetermined correction values. In such cases, the LTFT makes a more permanent correction to fuel delivery. In response, the STFT then accepts the new LTFT correction value as its 0 correction point. By way of example, if an engine has a small vacuum leak, the air-fuel mixture will be lean because too much air and insufficient fuel is present in the intake manifold. The STFT would temporarily adjust fuel delivery by adding more fuel to the engine in order to correct the problem. In this example (illustrated in Figure 2-10 on page 50), the STFT would add 6 percent more fuel (**Number 1** in graph). If the vacuum leak occurs over a long period of time, the LTFT would make a more permanent adjustment in fuel correction of +6 percent (**Number 2** in graph). Accepting this permanent adjustment, the STFT would then adjust its values again and use the LTFT +6 percent fuel correction number as its 0 point for all future fuel corrections (**Number 3** in graph)—or at least until LTFT adjusted value again. If the vacuum leak is large enough LTFT would be required to adjust its fuel delivery correction too far to make the proper fuel delivery to the engine possible, so the PCM would then set a P0173 (fuel trim malfunction) DTC and possibly turn on the MIL.

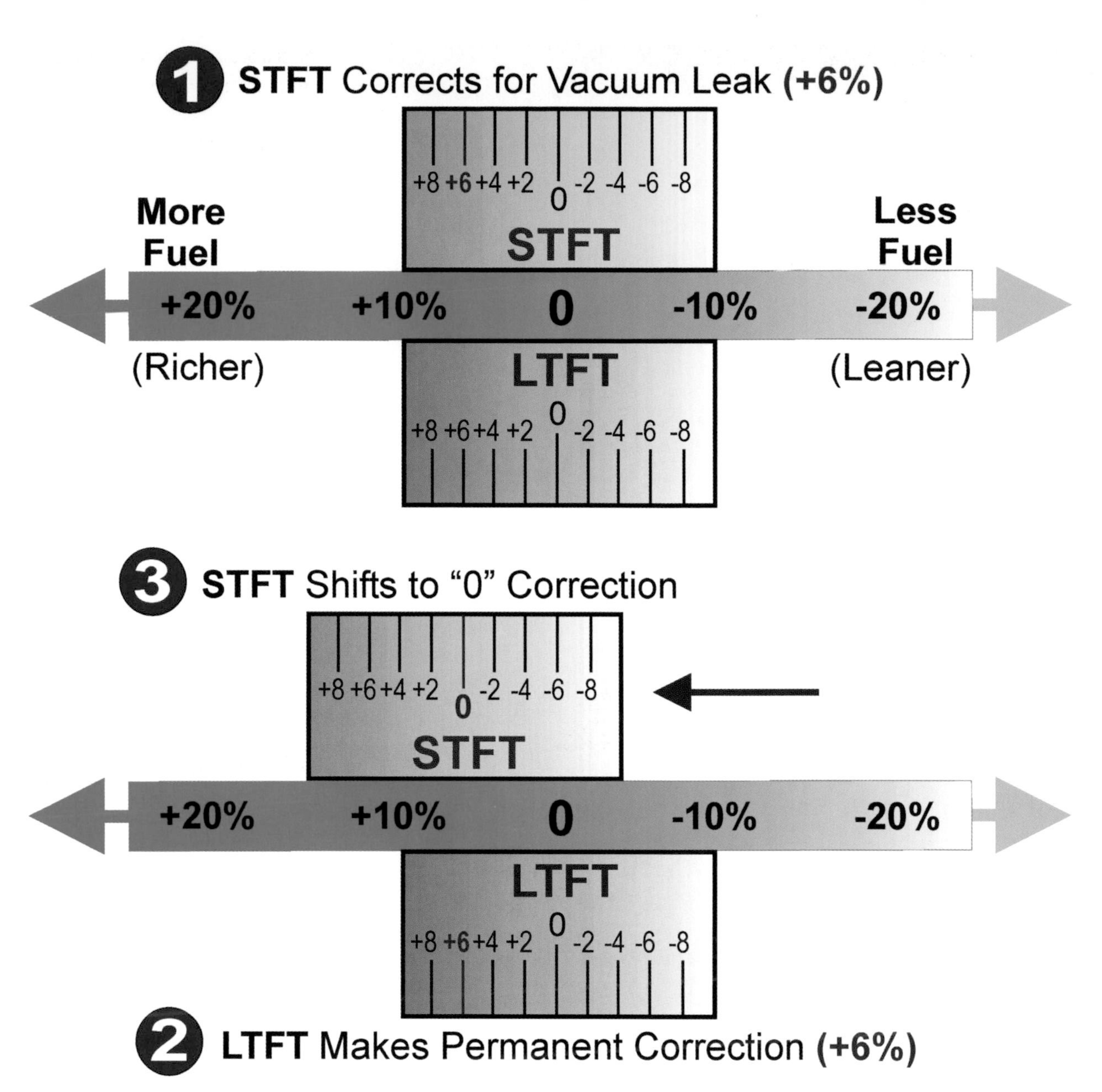

Fig 2-10—Whenever an engine is running, the short-term fuel trim (STFT) program is also operating, as is the long-term fuel trim (LTFT) program. STFT makes fuel corrections quickly, while LTFT makes fuel delivery corrections more slowly and determines its corrections based on the quicker STFT changes in fuel correction.

Oxygen Sensors and Heated Oxygen Sensor Monitors

Oxygen sensors provide major inputs to a PCM for adjusting air-fuel ratio for optimum catalytic converter efficiency. In addition to the fuel system, the EGR, catalyst, and EVAP systems all also rely on oxygen sensor input for their performance. OBD-II-equipped vehicles may use as many as four O_2 sensors. Because O_2 sensor performance affects so many other systems and all oxygen sensors must be hot in order to operate (600 degrees Fahrenheit), it's important these sensors start operating within a few minutes of start-up. All OBD-II system O_2 sensors use an electric heating element in addition to hot exhaust gas to accomplish rapid heating of the sensors.

An oxygen sensor monitor and heated oxygen sensor monitor are both two-trip monitors, meaning a PCM will watch for the occurrence of the same malfunction on two separate drive cycles before determining if a malfunction exists on an oxygen sensor. If a malfunction in an O_2 sensor is found during a first trip or drive cycle, the PCM will temporarily save the malfunction in its memory as a pending

All computer-controlled EGR valves regulate the flow of a small amount of exhaust gases into the intake manifold to reduce combustion temperatures in order to control the release of harmful oxides of nitrogen into the atmosphere. On General Motors' vehicles, the EGR valve is bolted to the intake manifold. The tube at the bottom of the valve (with the aluminum-colored, tinfoil-like wrapping) is connected directly to the exhaust manifold. The EGR valve is electrically operated by a vehicle's PCM, and opens only at part-throttle after the engine reaches normal operating temperature. *Courtesy Elwoods Auto Exchange*

code. The MIL does not come on at this time. If the same malfunction is again sensed during a second trip, the PCM will turn the MIL on and also permanently store a DTC code in its long-term memory. O_2 monitors check multiple operating systems for any of the following: heater functionality, too high or too low signal voltage, the rapidity with which signal voltage changes, an open or shorted O_2 circuit, and the length of time it takes for signal voltage to cross a 0.450 millivolt threshold (often referred to as cross counts).

Catalyst Efficiency Monitor and Heated Catalyst Monitor

No matter how efficiently an engine operates, a small amount of fuel always remains after the combustion process. A catalytic converter's job is to burn up this extra remaining raw fuel at the same time it is also eliminating or reducing other harmful emissions. It performs this process, in part, by storing oxygen and then adding it to the unburned fuel from the engine. The converter then chemically combines the stored oxygen and fuel so it is burned inside the converter.

A catalyst efficiency monitor (CEM) uses two oxygen sensors to determine if a catalyst is operating properly. One O_2 sensor, known as an "upstream" O_2 sensor (because of its upstream location between the engine and catalytic converter), determines the level of oxygen content in exhaust gases and then sends a signal communicating this level to the PCM for fuel delivery adjustments. A second O_2

sensor, commonly known as a "downstream" O_2 sensor (because of its downstream location between the catalytic converter and the exhaust's tailpipe), monitors a catalytic converter's ability to store oxygen. If a converter becomes damaged, and its ability to store oxygen compromised, a downstream O_2 sensor's signal will then match that of the upstream oxygen sensor's signal, which causes the converter to fail the monitor's test.

By contrast, a heated catalyst monitor tests a catalytic converter's ability to warm up quickly after the engine is started. The time it takes a converter to heat up is another indication of how well it is operating. Both a catalyst efficiency monitor and a heated catalyst monitor are two-trip monitors. As with O_2 sensors, if a malfunction is found during a first trip, the PCM temporarily saves the record of the malfunction in its memory as a pending code. The MIL does not come on at this time. If the same malfunction is again sensed during a second trip, the PCM turns the MIL on and also permanently stores a DTC code in its long-term memory.

EGR System Monitor

An exhaust gas recirculation (EGR) system redirects and adds a precise minimal amount of exhaust gases to the intake manifold while the engine is running at part throttle in order to reduce peak combustion temperatures to below 2,500 degrees Fahrenheit, so as to control and prevent the release of harmful oxides of nitrogen (NOx) gases into the atmosphere. An EGR system monitor checks the functionality of the EGR valve. There is no EGR sensor; consequently, a PCM uses several methods instead to monitor EGR flow, including: (1) a differential pressure feedback (DPFE) sensor that determines EGR flow by measuring pressure at two points between the exhaust and intake manifold; (2) an EGR temperature sensor, which measures temperature rise in the EGR passageway from the hot exhaust; and finally, (3) an EGR valve position sensor, which measures the amount or distance the valve opens. On certain vehicles, the PCM also monitors upstream oxygen sensor output voltage whenever the EGR valve is commanded to operate, as a means of determining if the EGR system is working properly.

Every time an engine is started on an OBD-II vehicle, an evaporative system monitor, or EVAP, checks the fuel tank venting system and gas cap for any leaks. If a gas cap is left open, off, or loose after adding gas to a fuel tank, the malfunction indicator lamp (MIL) will come on, and a DTC will be set. *Courtesy Elwoods Auto Exchange*

Pictured is a charcoal canister with various fuel venting hoses. This emission control device has been in use for a long time and serves as the main storage container for fuel vapors from the fuel tank. Once the engine is started and operational, the PCM will purge vapors from the canister at certain specified parameters and send them to the engine for subsequent burnoff. OBD-II systems are programmed to monitor both the fuel tank and vapor recovery system, including all of its various vapor hoses, for leaks every single time the engine is started. *Courtesy Elwoods Auto Exchange*

Like the sensors previously discussed, an EGR system monitor is also a two-trip monitor. Again, if a malfunction is found during a first trip, the PCM will temporarily save the malfunction in its memory as a pending code. If the same malfunction is again sensed during a second trip, the PCM will turn on the MIL as well as permanently save a DTC code in its long-term computer memory.

Evaporative System Monitor

Every OBD-II vehicle is equipped with an evaporative system monitor, or EVAP monitor, as it is more commonly known. This system monitor helps prevent fuel vapors from the fuel system from escaping into the atmosphere. An EVAP system routes gasoline fumes from the fuel tank into a charcoal canister storage container via hoses and tubes; the fumes are then stored in the canister and eventually rerouted back to the engine, where they are subsequently burned off in the combustion process. Switching valves and solenoids are used to accomplish the routing of these vapors within an EVAP system. In addition, an EVAP system monitor also tests all of the switches, valves, hoses, and the gas cap to ensure the EVAP system is leak-free. An EVAP system monitor also checks an EVAP system's ability to route fuel vapors on command. Any leakage in excess of a preset, predetermined amount will cause the system to fail an EVAP monitor's tests.

Like most of the sensors and systems previously discussed, an evaporative system monitor is also a two-trip monitor. After the second trip, the PCM turns on the MIL and also permanently sets a DTC code in its long-term memory bank. Furthermore, if a gas cap is left off, loose, or open, after filling the gas tank, the PCM will set a P0455 diagnostic trouble code defined as "Evaporative Emission Control System Leak Detected" (a gross leak). This DTC is accompanied by the malfunction indicator lamp on the dash lighting up to alert the driver to the need for repairs.

Secondary Air Injection Monitor

The use of air injection as a method for emissions control has also been around for a long time, and as a result of its efficiency, it is still in use today on some vehicles equipped with OBD-II systems. The process of air injection occurs when fresh air is injected by an air pump into the intake manifold or exhaust to assist with catalytic converter warmup when the engine is started cold. The additional air injected into the exhaust serves to add oxygen to exhaust gases, or in certain vehicles, directly into the catalytic converter for the same purpose.

A secondary air injection monitor uses an upstream oxygen sensor to determine if the air injection system is fully functioning. Like the majority of other monitors previously discussed, a secondary air injection monitor is also a two-trip monitor, which will require two complete drive cycles before the PCM will turn on an MIL or permanently store a DTC for a recurring malfunction that appeared in a system during both drive cycles. More information regarding the operation of electronic fuel injection systems, catalytic converters, exhaust gas regulating systems, and other emission control components and systems, are discussed in greater depth in Chapter 3, Electronic Fuel Delivery and Emissions.

CHAPTER 3
CATALYTIC CONVERTERS, OXYGEN SENSORS, AND ELECTRONIC FUEL DELIVERY

This chapter will focus on the way catalytic converters, oxygen sensors, and electronic fuel injection all function together in an OBD-II environment. A detailed look at the chemistry of what goes into an internal combustion engine, and what comes out as byproducts of the combustion process, will explain how a catalytic converter operates. We'll also explore the way the PCM, oxygen sensors, and the fuel injection system control the amount of fuel and oxygen in the exhaust gases and their subsequent effect on the catalytic converter's ability to function.

OBD-II and Catalytic Converters

A catalytic converter is the major player in the effort to reduce the bad stuff emanating from a vehicle's tailpipe. Understanding exactly what a "cat" does, and how it reduces exhaust emissions, will help when interpreting OBD-II catalytic converter-related DTCs and their causes. We'll start by exploring which chemicals are used by an internal combustion engine, how they are processed, what comes out of the engine's exhaust, and what happens inside a catalytic converter. Understanding the interrelationship of

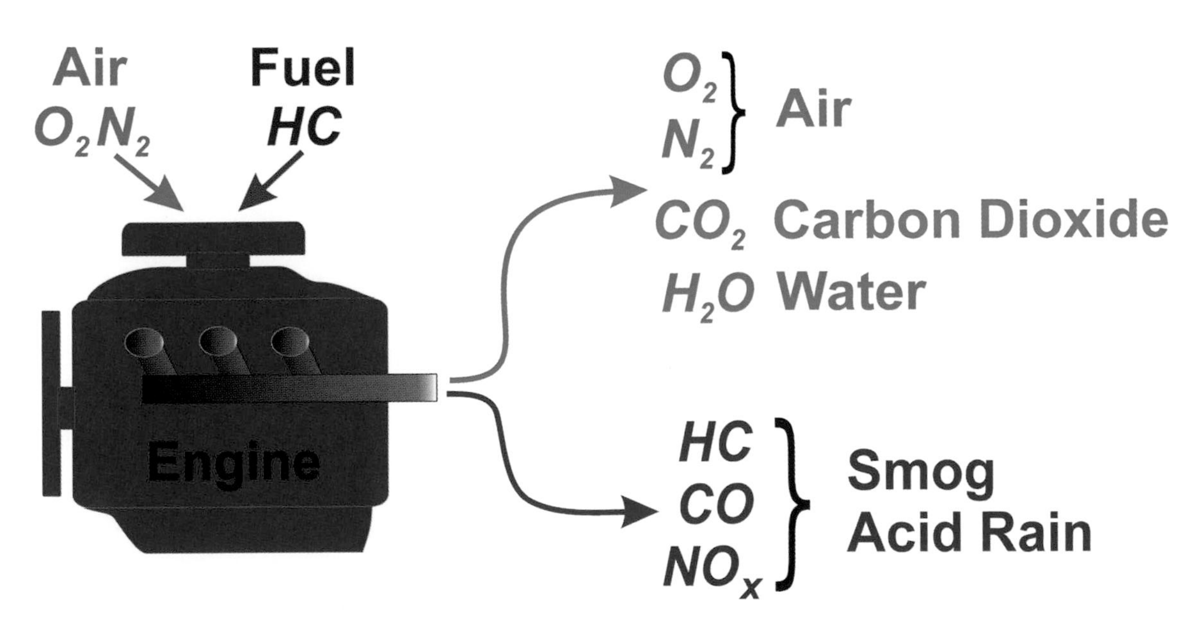

Fig 3-1—This diagram shows that when air (oxygen and nitrogen) is mixed with hydrocarbons in the form of gasoline, and then burned inside an engine, the resulting exhaust gases should consist of oxygen, nitrogen, carbon dioxide, and water (blue arrow). Unfortunately, even after more than 100 years of perfecting piston engine design, this process is still not as efficient as it could be, and consequently, pollution in the form of unburned fuel—or hydrocarbons (HC), carbon monoxide (CO), and nitrous oxide (NO_X)—is still emitted from exhaust gases (red arrow). Only the addition of a catalytic converter can eliminate the unwanted byproducts of combustion. Unfortunately, even with the addition of a catalytic converter, ozone-depleting CO_2 gases, which we now know are a major cause of global warming, are still an unavoidable byproduct of the combustion process.

OBD-II systems and catalytic converter functions will help with the task of interpreting 5-gas emission readings and emission failure test reports.

But first, a crash course in the chemistry of air and fuel and the dynamics of how they are processed within an internal combustion engine and catalytic converter is necessary. The atmosphere an internal combustion engine breathes is composed of 78 percent nitrogen and 21 percent oxygen. Nitrogen (N_2) is an inert gas, which means it simply takes up space and does not burn in the combustion process. Internal combustion engines create power from their ability to extract oxygen (O_2) from the atmosphere and combine it with hydrocarbons from gasoline, and then burn the combined air/fuel mixture. The combustion process basically creates heat, which causes nitrogen gases inside the combustion chamber to expand and push the piston down the cylinder, which then rotates the engine's crankshaft. Engines with high compression ratios, long duration camshafts, superchargers, turbochargers, and large displacement engines all create more power, because they convert greater amounts of fuel and oxygen into heat.

Gasoline is made mostly from hydrocarbons (HC), which, when combined with oxygen (O_2) and ignited, produce heat. If piston engines were 100 percent efficient, all of the HC and O_2 would be burned in the combustion process, and only water (hydrogen bonded with oxygen—or H_2O), nitrogen (N_2), and carbon dioxide (CO_2) would be left. Unfortunately, the combustion process is not 100 percent efficient, and harmful unwanted gases always remain. These chemical "leftovers" are emitted from a vehicle as exhaust emissions in the form of three hazardous gases: carbon monoxide, hydrocarbons, and oxides of nitrogen (NO_x). The first of these three harmful gases is carbon monoxide (CO), which is colorless, odorless, and poisonous. Carbon monoxide is what kills people when they leave a car running inside a closed garage. The next unwanted byproduct of combustion is unburned fuel or HC (hydrocarbons), also called VOC for volatile organic compounds. The last unwanted gas is known as oxides of nitrogen, or NO_x, as it is commonly called. When combined with HC, NO_x produces smog, which we all know makes people's eyes sting on a hot summer day and makes it hard for all living things to breathe. No matter

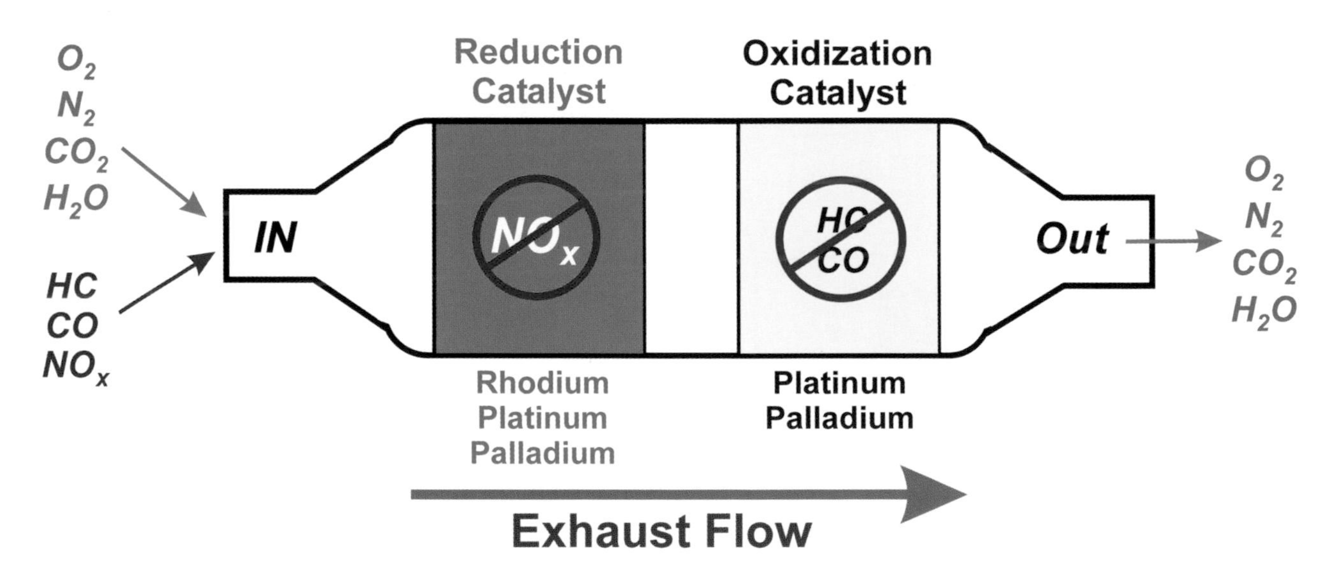

Fig 3-2—Catalytic converters used on both OBD-I and OBD-II systems are referred to as three-way converters because they reduce or eliminate three bad gases coming from the engine: HC, CO, and NO_x. There are two stages during which reduction and elimination of harmful gases from exhaust emissions are accomplished by a catalytic converter. The first processing stage of a three-way catalytic converter is known as the reduction catalyst stage; this stage eliminates NO_x. All, or nearly all, HC and CO gases are consumed in the oxidization catalyst during the second processing stage of the converter. Some converter designs involve the use of a tube that supplies additional air (oxygen) to the center of the cat to aid the oxidization process. Fortunately, all of the gases that exit the catalytic converter are safe to breathe and don't make smog. When the converter is operating efficiently, the only gases exiting the tailpipe are oxygen, nitrogen, carbon dioxide, and water vapor.

how high-tech the design, all piston engines produce these pollutants.

Catalytic converters, or "cats," have been equipped on some cars since 1975. Modern converters are constructed using an extruded honeycomb structure or core. The core, or substrate, is sprayed or dipped and then subsequently coated with precious metals, including platinum, palladium, and rhodium. Platinum and palladium accelerate the oxidation of HC and CO so they won't exit the tailpipe and pollute the atmosphere. The use of rhodium helps reduce NO_x emissions.

In order for a catalytic converter to start the process of converting harmful exhaust emissions into more desirable exhaust compounds, the cat must be hot, around 500 degrees Fahrenheit, which is known as the "light-off" temperature; normal operating temperatures for catalytic converters range between 750 and 1,500 degrees Fahrenheit or higher. A cat should normally last the life span of an engine. However, if the fuel system consistently runs an air/fuel mixture that is too rich, or an ignition system continually misfires, the temperatures within the "cat" can increase to more than 2,500 degrees Fahrenheit, which will shorten its lifespan or destroy it. Other factors also affecting converter longevity are the use of leaded fuel, excessive engine oil burning, or the presence of antifreeze in the exhaust from a blown head gasket or cracked cylinder head.

Converting Bad into Good

Early catalytic converters in use in the mid-1970s were two-way, meaning they were only capable of reducing two harmful gases from exhaust emissions—hydrocarbons (HC) and carbon monoxide (CO); unfortunately, harmful NO_x gases were not eliminated by these early model cats. However, since the 1980s, three-way catalytic converters have now been the norm, and fortunately, they do effectively reduce all three of the harmful gases contained in exhaust emissions—namely, HC, CO, and NO_x gases.

A three-way catalytic converter removes or eliminates all three harmful gases from exhaust emissions during a two-stage process that consists of two separate catalyst stages—reduction and oxidization. This two-stage design is also known as a dual bed converter. The reduction catalyst processing stage is the first stage that all exhaust gases must pass through, and its function is to ideally eliminate, or at least reduce, NO_x emissions. An NO_x molecule is composed of one atom of nitrogen and several oxygen atoms. By using the elements platinum and rhodium, the reduction catalyst rips the nitrogen atom out of the molecule, thus freeing the oxygen. The free nitrogen atoms bond with each other to create N_2 (essentially nitrogen), which fortunately makes up 78 percent of the earth's atmosphere. The free O_2 is then passed into the second stage, or oxidization stage, of the catalytic converter where it will be put to good use. The second processing stage, or oxidization catalyst stage, reduces HC and CO from exhaust emissions by adding the oxygen freed from the NO_x molecule during the reduction processing stage of the cat, with excess oxygen from exhaust gas. Based on input from an oxygen sensor, the PCM attempts to maintain an ideal average air/fuel ratio of 14.7:1. Thus, the fuel injection system constantly varies the air/fuel mixture ratio between just slightly rich, and slightly lean, of the ideal average 14.7:1 air/fuel mixture ratio. This variation in air/fuel mixture allows enough oxygen to exist in exhaust gases for the oxidization catalyst stage of the catalytic converter to convert HC and CO into harmless water and less harmful carbon dioxide (two compounds naturally found in the earth's atmosphere). Certain catalytic converter designs require more oxygen than is normally produced by a standard cat design, so these special cats use an external air source and a pump in order to pump air into the center of the cat so there is sufficient oxygen present in the oxidization catalyst stage to perform harmful emissions removal processes. The gases coming out of a catalytic converter are so clean that when sniffed by a 5-gas analyzer, typical readings should register "0" detected presence of HC, CO, and NO_x. In fact, a cat is so effective in cleaning exhaust gases, the rate of suicide from breathing engine exhaust has dropped significantly since the introduction of catalytic converters on vehicles.

It is interesting to note that all through the 1980s, and midway through the 1990s, automotive textbooks and service manuals claimed that a catalytic converter's ability to eliminate HC, CO, and NO_x, was sufficient to completely rid the earth of pollutants from automobiles. While this might be true to a limited extent, it wasn't until the late 1990s, and especially now in the new millennium, that humans have realized that simple reduction of harmful chemicals from automobile emissions is not sufficient to solve the problem of global warming. Today, most scientists worldwide agree that manmade CO_2 emissions are the primary cause of rising temperatures on earth. Ironically, the higher the CO_2 levels an internal combustion engine produces, the more efficient it is at burning fuel, but unfortunately, the larger the engine, the greater amount of CO_2 it actually produces. The use of smaller, more fuel-efficient engines on vehicles, as well as hybrid engines that use a combination of battery power and power extracted from hydrocarbons (gasoline), looks to be more promising for effective pollution control, at least for the immediate future with regard to automobile production.

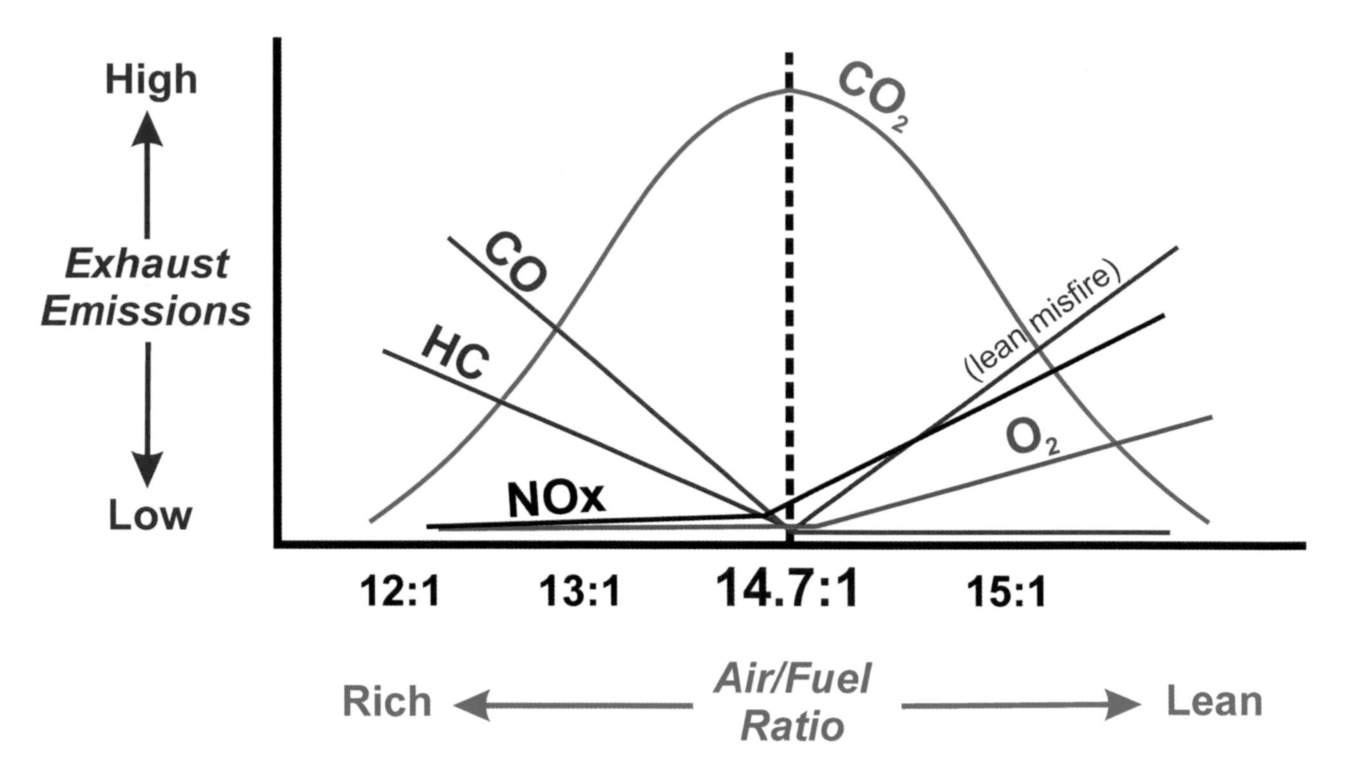

Fig 3-3—The graph in Figure 3-3 provides a quick reference for interpreting exhaust gas readings produced by a five-gas analyzer. A 14.7:1 air/fuel ratio is ideal for combustion since harmful, unwanted gases such as HC, CO, and NO_x are at their lowest levels achievable with current mainstream technology. As the air/fuel ratio inclines toward the rich side of ideal, both HC and CO start to rise. When the air/fuel ratio indicates a rich fuel mixture, combustion temperatures are relatively low, and NO_x production will be zero. When the air/fuel mixture tends toward the lean side of ideal, CO is zero, because enough oxygen is present for complete combustion. However, if the mixture is more lean than desirable, too much oxygen and not enough fuel will be present in the mixture, causing a lean misfire to occur. A lean misfire occurs whenever there is insufficient fuel present for combustion to occur. Lean misfires cause extremely high levels of HC (raw unburned fuel) to exit the cylinder, since only limited amounts of fuel were present that were not ignited by the spark plug.

Air/Fuel Ratios and 5-Gas Exhaust Analyzers

Exhaust gas analyzers (EGA) have been around for years. These diagnostic tools can measure two, three, four, or five exhaust gases, depending on the sophistication level of the model utilized. While the cost of these tools (typically upwards of $3,000) is customarily beyond the budgets of most shade tree mechanics, knowing how they are used by professional technicians is helpful to understand and interpret readings from exhaust gases contained in emissions' failure reports or work orders from repair shops or new car dealers.

The 5-Gas Air/Fuel Ratio graph shown in Figure 3-3 provides the easiest explanation of the relationships between exhaust gases, as well as their relationship with air/fuel mixtures and how these mixtures affect them. An ideal average air/fuel ratio of 14.7:1 produces the most efficient and complete burning of the air/fuel mixture during combustion. This ratio is the ideal ratio that an EFI system averages at idle or cruise in "closed loop" operational mode. The chart in Figure 3-3 illustrates why the ideal air/fuel ratio of 14.7:1 produces the lowest level emissions of HC, CO, and NO_x gases. As an air/fuel mixture gravitates toward the rich side of the ideal 14.7:1 ratio, both hydrocarbons (HC) and carbon monoxide (CO) emissions start to rise. Since CO is a byproduct of incomplete combustion, it is only produced when insufficient oxygen is present in the air/fuel mixture. By contrast, the presence of CO in exhaust emissions indicates a rich air/fuel mixture—consequently, if exhaust gas readings detect CO in exhaust gas, the air/fuel mixture must necessarily be rich (more fuel, less air). Lack of oxygen in a rich mixture will also cause HC to rise, since not all of the fuel can be burned in that instance as well.

So far we've only looked at three bad gases: HC, CO, and NO_x. Now let's take a look at carbon dioxide (CO_2) emissions. CO_2 is often overlooked as an exhaust gas, but it does have value for air/fuel mixture fine-tuning. CO_2 readings peak in output at the same average ideal 14.7:1 air/fuel ratio. The higher the CO_2 percentage present, the more efficiently the engine is combusting air and fuel. Since CO_2 readings flatten out at peak, it can often be difficult to tell on which side of the ideal 14.7:1 air/fuel ratio (either rich or lean) the air/fuel mixture is at. Levels of carbon monoxide (CO), are required to determine when CO_2 is at its highest point (which is the same point the engine is also at its most efficient). CO_2 is at its maximum when CO readings first drop to 0 percent. By looking at CO percentages in conjunction with CO_2 percentages, peak CO_2 can be determined. CO_2 levels can tell you how efficient an engine is and how close to ideal its fuel mixture is.

In lean air/fuel mixtures, or those that tend to the lean side of the ideal 14.7:1 ratio, CO is close to or at 0 percent, because sufficient oxygen is present in the exhaust gases to support complete combustion. Thus, O_2 is also a lean air/fuel mixture indicator, since if an O_2 reading detects the presence of O_2 in exhaust gases, the air/fuel mixture must necessarily be lean. However, this is not always the case. A false lean air/fuel mixture reading can be caused by the presence of a malfunctioning air injection system or by leaks in an exhaust system. As the air/fuel mixture continues to lean out further (less fuel, more air), the molecules of hydrocarbons (HC) become spaced further apart in relation to the molecules of air, rendering combustion via ignition from a spark plug that much more difficult. As HC molecules become even more widely spaced, because less fuel is present, a lean misfire will occur once the spark plug fails to ignite the air/fuel mixture. When this occurs, raw, unburned fuel passes directly into exhaust gases, causing HC readings to skyrocket.

In addition, NO_x emissions also increase whenever air/fuel mixtures are on the lean side of ideal and there is excess oxygen present during combustion. A lean mixture burns hotter than a rich air/fuel mixture, and as a result, increased temperatures produce NO_x as a byproduct of the combustion process. The engine's exhaust gas recirculation (EGR) system subsequently reduces the presence of NO_x by adding inert exhaust gas (in the form of nitrogen) back into the intake manifold. The inert nitrogen gas serves to reduce combustion temperatures from lean air/fuel mixtures, thereby reducing NO_x gas levels emitted during combustion.

The 5-gas readings illustrated in Figure 3-3 represent readings obtained from exhaust gases coming directly from the engine. Exhaust gas readings sampled at locations after a catalytic converter, or at the tailpipe itself, will be "masked" by the operation of the catalytic converter. This is because the converter's job is to eliminate HC, CO, and NO_x gases from exhaust emissions, so that only small amounts of O_2 and CO_2 will be present instead in the exhaust.

The five-gas exhaust analyzer shown in the photograph measures the amount of hydrocarbons (HC), carbon dioxide (CO_2), oxygen (O_2), carbon monoxide (CO), and oxides of nitrogen (NO_x), present during the combustion process. In view of the fact an OBD-II system can itself indirectly determine if an engine is producing excessive levels of HC, CO, or NO_x simply by monitoring a catalytic converter's performance, five-gas analyzers are no longer used as frequently as they once were, except perhaps for emission repairs' verification or to perform basic engine diagnostic work. *Courtesy Keplinger's Automotive Center*

Besides utilizing a 5-gas analyzer exhaust reading to determine if an engine is running with a rich or a lean air/fuel mixture, an exhaust gas analyzer can also uncover other mechanical engine-related problems. By holding an analyzer's sampling probe just above the level of coolant present in a radiator (naturally, with the radiator cap removed), any reading that detects the presence of hydrocarbons (HC) could indicate a possible blown head gasket or cracked cylinder head may have occurred. Technicians must be careful performing this test, since if they suck any coolant up into the probe, the 5-gas analyzer test machine will be completely ruined. Another different mechanical condition test can also be performed by inserting a five-gas analyzer's sampling probe into a crankcase breather hose; high HC readings from this test can indicate worn piston rings or other cylinder sealing problems.

A 5-gas analyzer tool can also be used to perform a cylinder power balance test. By shorting individual spark plugs to ground, and then taking a gas reading, the efficiency with which each cylinder combusts the air/fuel mixture can be revealed. Each time a cylinder is "shut off" (no spark), HC readings will increase and CO_2 will drop. Comparison of each nonoperating cylinder will reveal how hard each one is working. If one cylinder is performing less optimally than others (its CO_2 reading does not drop as much as other, stronger cylinders), this test will confirm that cylinder was not pulling its weight (so to speak), and a possible mechanical problem may exist. Fuel injectors can additionally be tested for optimal fuel atomization and spray pattern. HC readings should increase the same amount for each cylinder as each cylinder's corresponding spark plug is shorted to ground and the cylinder has stopped running. Consistent HC readings confirm all fuel injectors are spraying the same amount of fuel into the engine.

All About Oxygen Sensors

So far, we have explored the chemistry of air and fuel as they are consumed by the engine and how a catalytic converter operates. It's important to keep in mind the reason we have EFI and OBD-II engine management systems at all is to provide proper air/fuel ratios to keep catalytic converters operating at peak efficiency at all times. The key to having precisely the correct amount of air and fuel burning inside a combustion chamber and then exiting the engine as exhaust gas, and eventually passing through a catalytic converter, is a properly functioning oxygen sensor. Earlier OBD-I systems used only one or two O_2 sensors, located between the engine and a catalytic converter. Oxygen

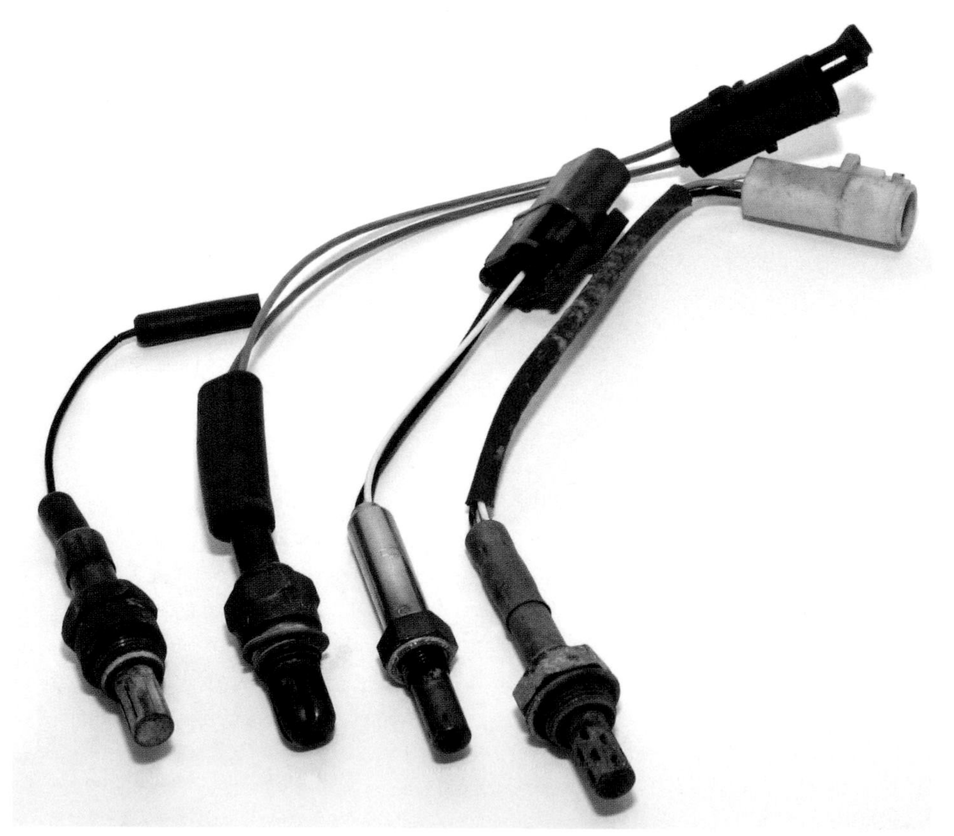

In the photo are four O_2 sensors with different numbers of wires coming from them. All modern OBD-II systems use oxygen sensors fitted with internal heaters. The internal heating element keeps the sensor at ideal operating temperature, even if exhaust gas temperature drops low due to prolonged engine idling. In addition, the heating element immediately gets the sensor hot enough to operate whenever an engine is started cold, so the system can operate in closed loop mode more quickly. The number of wires an oxygen sensor has can be used to identify if the sensor uses an internal heating element. Oxygen sensors with one or two wires don't have internal heaters and are typically found only on older OBD-I systems. Oxygen sensors with three or four wires have internal heating elements and are found on certain vehicles equipped with OBD-I systems and all modern OBD-II systems.

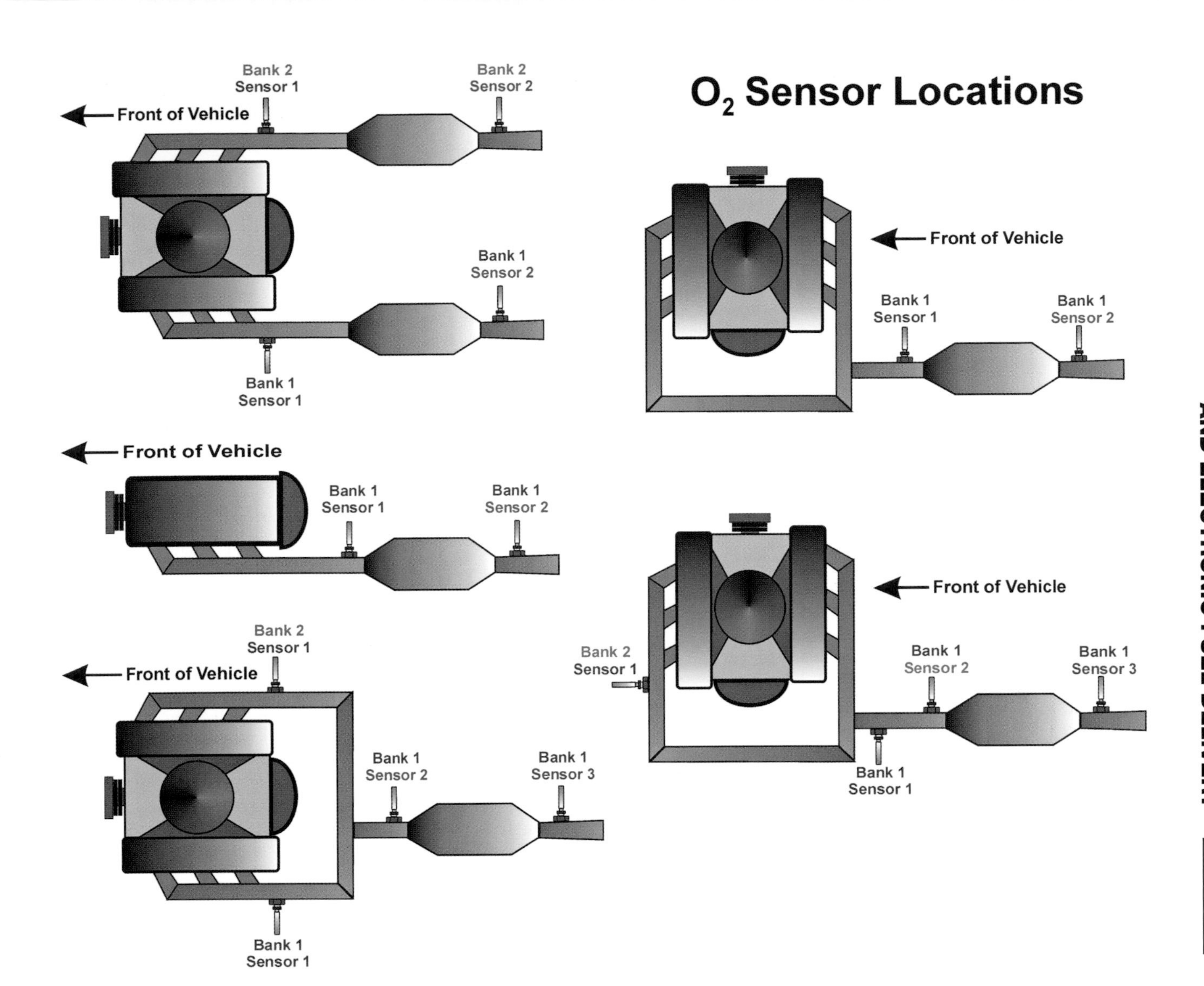

Fig 3-4—Since auto manufacturers can locate oxygen sensors wherever they want, it can often be confusing to identify which O_2 sensor one is dealing with when viewing oxygen sensor data on a scan tool. It is not uncommon to replace a wrong O_2 sensor simply because of difficulty identifying sensor locations. This diagram shows different engine configurations and layouts of exhaust systems, as well as their related upstream and downstream oxygen sensors in relation to a catalytic converter. By way of illustration, the engine depicted top left uses four oxygen sensors: Bank 1/Sensor 1, and Bank 1/Sensor 2 are used on the right side of the engine; Bank 2/Sensor 1, and Bank 2/Sensor 2 are located on the other side of the engine and comprise the other bank of cylinders on this V engine type. Other converter/oxygen sensor layouts are shown as well.

sensor(s) on these early emission systems provided information the computer used to constantly correct the air/fuel ratio to match a converter's requirements for reducing exhaust emissions. More modern OBD-II systems use one or more O_2 sensors, also located between the cat and the engine, plus an additional O_2 sensor, or sensors, located between a catalytic converter and the tailpipe. In OBD-II systems, the first oxygen sensor (or set of sensors) located between the engine and catalytic converter is/are known as the "upstream" sensor(s), while the O_2 sensor(s) located between the converter and the tail pipe is/are known as the "downstream" sensor(s). The upstream sensor(s) next to the engine have nothing to do with the downstream oxygen sensors.

All OBD-II systems have a minimum of two, and as many as four, oxygen sensors. Essentially there will always be an O_2 sensor located between the engine and the catalytic converter, and another O_2 sensor located between the converter and the exhaust tailpipe. This is true even when two catalytic converters are used (see Figure 3-4 on page 61).

We'll start with a discussion of the two types of commonly used O_2 sensors equipped in vehicles today—zirconia and titania. Later, more in-depth coverage will be provided to upstream and downstream oxygen sensors.

The first type of oxygen sensor is a zirconia dioxide sensor that acts as a galvanic battery by comparing oxygen content inside the exhaust to the oxygen content in the surrounding atmosphere in order to generate a small, variable DC voltage. This voltage is interpreted by a PCM as a signal that represents an air/fuel ratio which may be ideal, too lean (too much oxygen/too little fuel), or too rich (too much fuel/too little oxygen). Here's how it works: When oxygen content in the exhaust gas is low (a rich air/fuel mixture), the difference between the level of exhaust gas oxygen and the level of oxygen in the atmosphere is high, causing the sensor to produce a relatively high voltage—between 0.5 to 0.9 volt. Conversely, when the O_2 sensor detects a high exhaust gas oxygen content (a lean air/fuel mixture), and compares it with the oxygen in the outside atmosphere, the difference is smaller, so consequently, lower voltages are generated—between 0.1 and 0.4 volt. The output voltage is then interpreted by a PCM as either ideal, too much fuel (rich), or too little fuel (lean).

The other type of oxygen sensor is a titania O_2 sensor; it operates somewhat differently from a zirconia type sensor, but the end results are the same. Instead of producing a small voltage, a titania O_2 sensor uses a reference voltage from a PCM and then adjusts its internal resistance based on the oxygen content found in the exhaust gas. A titania sensor's resulting voltages are similar to those of a zirconia sensor. A rich fuel mixture produces higher voltages (above 0.45 volts), while a lean fuel mixture with too little fuel produces lower voltages of less than 0.45.

Both oxygen sensor types must be hot (600 degrees Fahrenheit), before they can function. Oxygen sensors have to rely on hot exhaust gases to maintain optimal operating temperature levels. Consequently, all modern OBD-II sensors use an internal electric heating element to prevent them from cooling down at idle (when engine exhaust gas temperature is low), and to facilitate a faster sensor warm-up time during cold starting.

Upstream O_2 Sensor

An upstream O_2 sensor operates in the same manner and performs the same function in both earlier OBD-I systems and in more modern OBD-II systems. The upstream O_2 sensor is responsible for closed loop operation, in conjunction with a PCM and the fuel injection system. Operation of an upstream O_2 sensor can be observed via a scan tool that displays engine data streams or via use of a digital

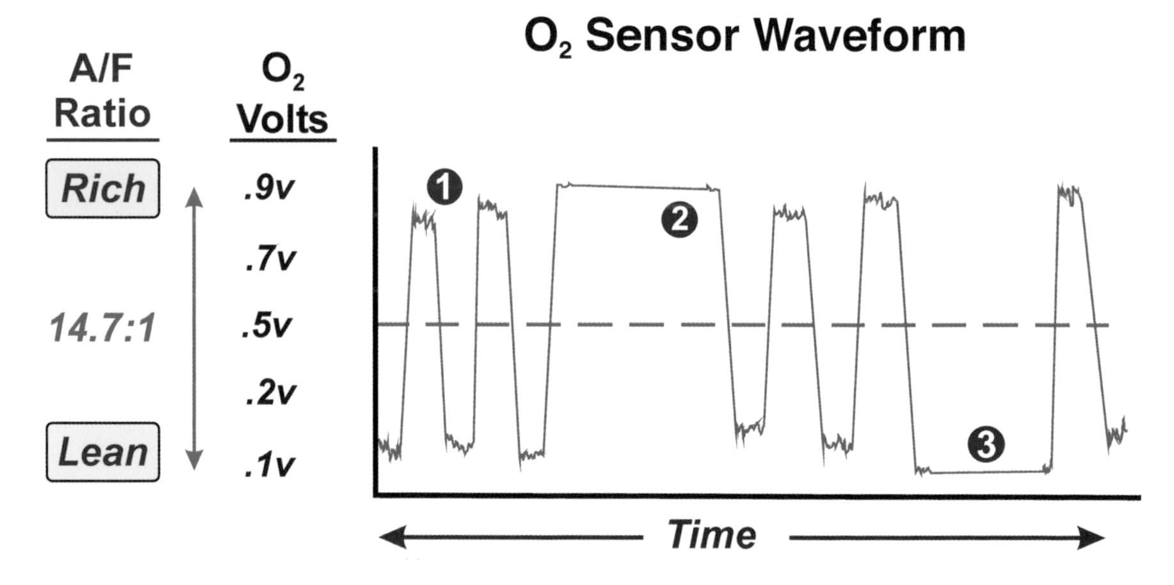

Fig 3-5—This graph shows an oxygen sensor's waveform as displayed on a lab oscilloscope. In the illustration, the engine is at 2,000 rpm with a steady throttle. On the graph, position number 1 indicates the O_2 sensor is operating in closed loop mode because voltage keeps fluctuating up and down (reflecting continuous fuel correction adjustments made by a PCM), around an average of about 0.5 volts. When the throttle is snapped open (position number 2 on the graph), O_2 sensor voltage rises to 0.9 volts, indicating a rich air/fuel mixture. Between position numbers 2 and 3 on the graph, the fuel system returns to closed loop operational mode. Position number 3 on the graph indicates a lean air/fuel mixture when the throttle is suddenly closed. This lab scope waveform confirms the O_2 sensor is working correctly. Certain scan tools can display oxygen sensor output voltages in graphic formats that look similar to the one displayed here.

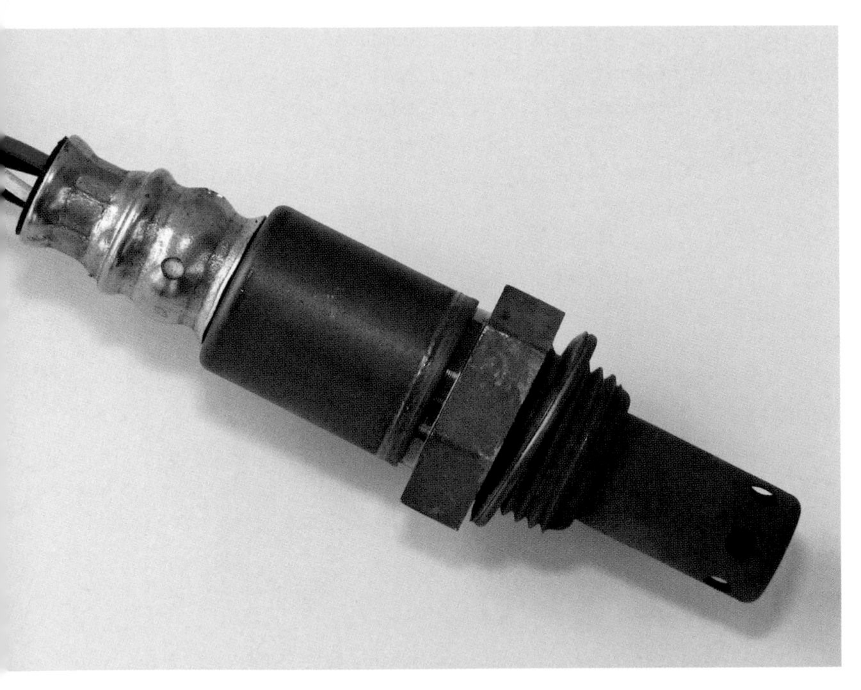

Many manufacturers use the same type of O_2 sensor to sense oxygen content in exhaust gases, whether the location of the sensor(s) is in front of the catalytic converter (between the engine and the cat) or behind the catalytic converter (between the cat and the exhaust tailpipe); these sensors are also commonly referred to as upstream and downstream O_2 sensors, respectively. Early OBD-II-compliant vehicles often used oxygen sensors that were interchangeable between upstream and downstream locations. Newer OBD-II-compliant vehicles are instead equipped with wiring harnesses of different lengths to accomplish the same function, some of which have different connectors, making it impossible to plug a rear downstream O_2 sensor into a front upstream O_2 sensor harness. The late-model Toyota O_2 sensor in the photo may look exactly the same for front upstream and rear downstream applications, but the connectors are different, making interchangeability difficult. *Courtesy Younger Toyota*

This 2.4-liter, four-cylinder engine for a Toyota RAV4 uses two upstream oxygen sensors (yellow arrows). One sensor monitors exhaust gas from cylinders 1 and 2, while the other monitors similar exhaust gases from cylinders 3 and 4. Even though these O_2 sensors are located inside the exhaust manifold, they still use their own internal electric heaters to quickly drive up their temperatures to minimum operating levels when the engine is cold. *Courtesy Younger Toyota*

voltmeter connected directly to an oxygen sensor's output signal wire. By monitoring O_2 sensor voltage, it's fairly easy for a PCM or scanner tool to determine if a sensor is indeed working, and if so, how well it's working.

To test an O_2 sensor, the engine should be warmed up to normal operating temperature by driving the vehicle, rather than merely letting it idle for a few minutes. Set up a scan tool to enable it to read an upstream O_2 sensor (some vehicles have more than one), or use a digital voltmeter connected to an oxygen sensor. Start the engine and maintain speed at 2,000 rpm for 60 seconds. If O_2 voltage starts fluctuating back and forth, somewhere between 0.2 and 0.8 volt, the EFI system is operating in closed-loop mode. (However, bear in mind that for purposes of this test, it doesn't really matter if the system is in closed-loop or open-loop operational mode). Next, while watching the scan tool

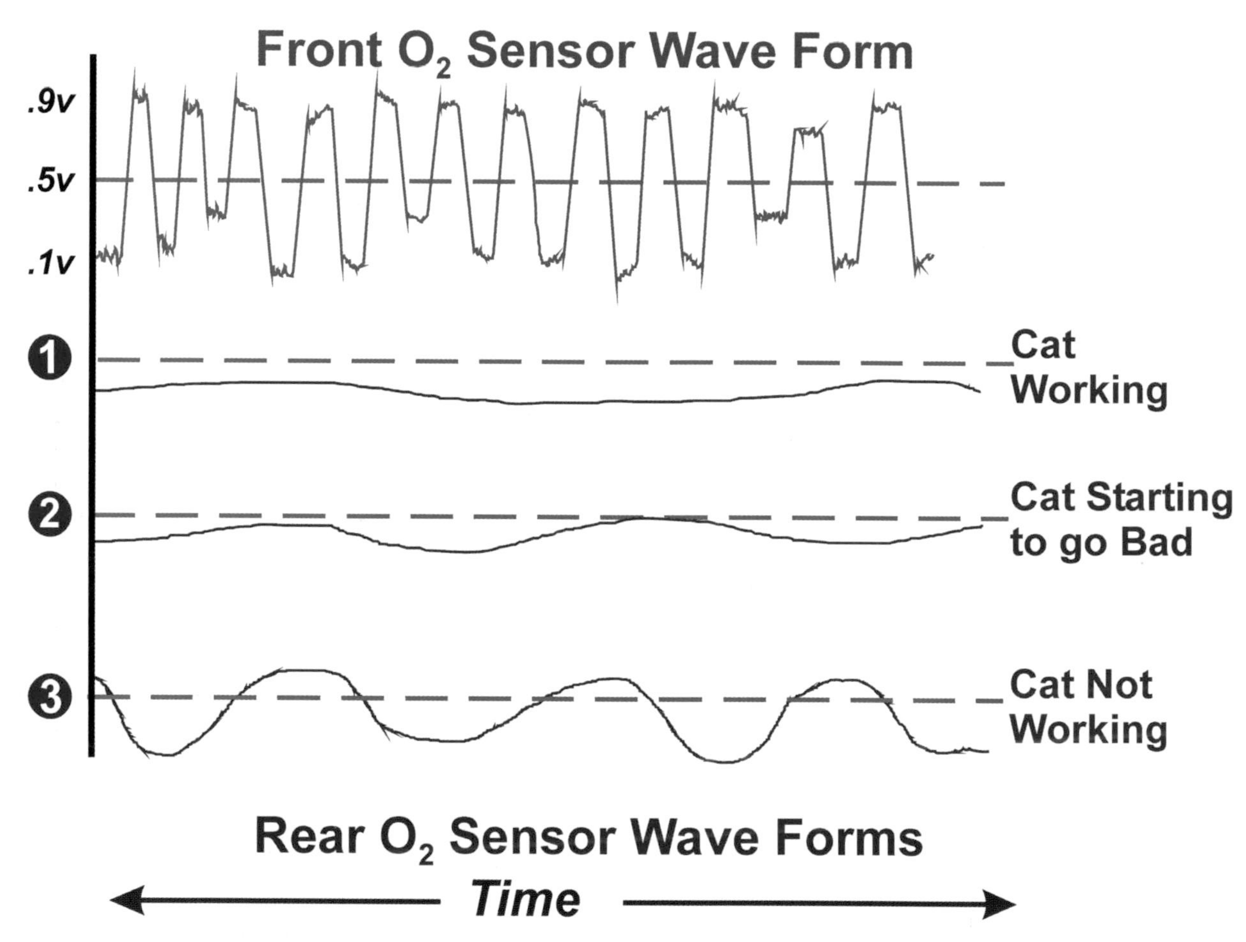

Fig 3-6—The lab oscilloscope waveforms in the graph represent O_2 sensor voltage output over time. (Voltage is shown on the vertical scale, while time is shown on the horizontal scale.) Viewing oxygen sensor output voltages in a graphic format allows technicians to observe the quality of a sensor's output signals. The waveform in blue (top) shows the operation of a front upstream (before the converter) O_2 sensor, as it alternates between a lean and rich mixture during closed loop operation. As depicted in the graph, waveform No. 1 represents the waveform for the rear downstream (after the converter) O_2 sensor during the same time period as front upstream O_2 sensor operation. The rear downstream sensor's voltage signal confirms the catalytic converter is operating normally. No. 1 voltage waveform only changes slightly, and much slower than the upstream oxygen sensor and illustrates the catalysts ability to store and release oxygen. Also depicted in the graph, waveform 2 represents a catalytic converter starting to lose its ability to store oxygen, as the rear downstream O_2 sensor is starting to look similar to the voltage output of the front upstream O_2 sensor. The waveform in No. 2 is alternating between high and low voltages readings and it is changing more rapidly than the waveform No. 1. In the graph, waveform 3 shows the catalytic converter is completely worn out and can no longer store and release enough oxygen to reduce exhaust emissions—in effect the cat is chemically not there; therefore, the downstream O_2 sensor looks more similar to the upstream O_2 because both voltage signal waveforms look more alike. When the upstream and downstream O_2 sensor waveforms look too much alike (electrically speaking) the PCM will set a DTC that indicates the catalytic converter is not functioning.

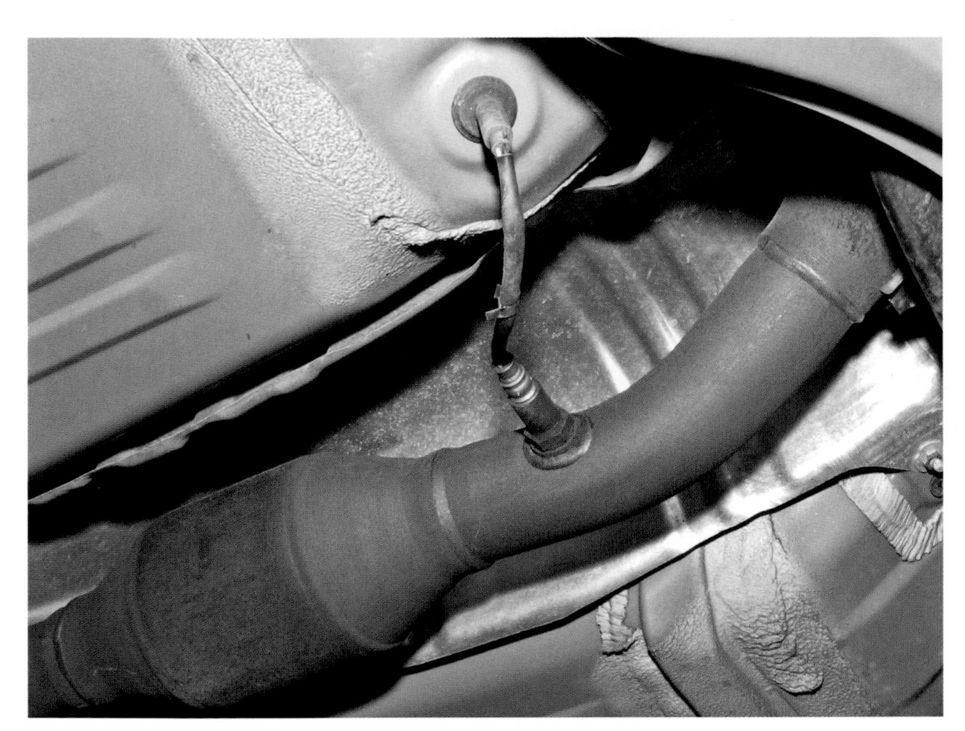

This downstream oxygen sensor is about a foot away from the cat (not shown), and located to its right, up next to the engine. On this Toyota, the downstream O_2 sensor is basically identical to the upstream oxygen sensor; in fact, on this vehicle, the wiring harness length and connector are the only differences between the upstream and downstream O_2 sensors. *Courtesy Younger Toyota*

or voltmeter, snap the throttle open. O_2 voltage should increase to 0.9 volt, indicating a rich fuel mixture. Normally, there is a slight delay from the time the throttle is opened until the time a scan tool is able to display an increase in voltage. This delay should be taken into account when using a scan tool for oxygen sensor testing. When using a digital voltmeter, changes in O_2 sensor voltage should be instantaneous. Then, hold engine speed steady again at 2,000 rpm, and then quickly close the throttle. This time, O_2 voltage should drop to 0.1 volt or less, since the PCM has cut off fuel to the engine in order to create a leaner mixture. How fast an O_2 sensor responds to fluctuations in exhaust gas oxygen, as well as the range of voltage displayed (0.1 to 0.9 volt), indicates whether the sensor is good or bad. A good sensor should be able to make voltage transitions instantly (a digital voltmeter connected directly to an O_2 sensor's signal wire is usually much quicker at displaying results than a scan tool), while a lazy or worn-out O_2 sensor will only make slow voltage transitions; it also won't be able to reach 0.9 volt, no matter how rich the fuel mixture.

Downstream O_2 Sensor

The main function of a downstream O_2 sensor is to monitor a catalytic converter's performance, to ensure it's functioning properly. In view of its function, a downstream O_2 sensor is also sometimes called a "cat monitor oxygen sensor." To determine a catalytic converter's efficiency, a PCM compares voltage signals from the upstream and downstream O_2 sensors. In closed-loop operational mode, the fuel system alternates the air/fuel ratio between a slightly rich mixture and a slightly lean mixture. This air/fuel ratio supplies the catalytic converter with oxygen, so that when the air/fuel mixture is leaned out (more air added), the cat will store oxygen to reduce emissions. When the fuel mixture is adjusted to a richer mix, the cat releases oxygen into the exhaust gas, and then uses it to convert hydrocarbons (raw fuel) and carbon monoxide (a harmful byproduct of combustion) into water vapor and carbon dioxide, thus eliminating, or at least reducing, noxious exhaust emissions into the atmosphere.

A catalytic converter's ability to store and release oxygen is an indication of how healthy it is. When a cat stores oxygen as a result of a lean air/fuel mixture, oxygen is low (as sensed by a downstream O_2 sensor), thus resulting in a fairly constant output voltage signal By contrast, when the air/fuel mixture is rich, the stored oxygen inside the cat is released, and again, the oxygen exiting the cat will only change an incremental amount, causing the downstream O_2 sensor to register only minimal voltage changes.

Even when brand-new, a catalytic converter's ability to store oxygen is limited. When the maximum available storage limit of a converter is reached, a downstream O_2 sensor will indicate either a high oxygen content (indicated by low voltage on a scan tool) or low oxygen content (high voltage displayed on a scan tool). During normal operation, the

Older OBD-I systems were not designed to monitor the performance of a catalytic converter. In fact, a converter could be damaged or even completely missing, and an OBD-I diagnostic system would not even turn on a check engine MIL to warn a driver that the vehicle was puking out pollutants. Only with the introduction of modern OBD-II diagnostic systems has a converter's performance been able to be monitored as part of onboard diagnostics. In addition, the PCM on an OBD-II system will flash a service engine soon MIL if the converter is being damaged by an ignition misfire. *Courtesy Kiplinger's Automotive Center*

In addition to running the electronic engine management system and OBD-II emissions diagnostics, a PCM (shown in photo) also shifts the automatic transmission and manages body control functions, such as power door locks, vehicle alarm, power windows, and climate control. *Courtesy Elwoods Auto Exchange*

downstream O_2 sensor's high and low voltages are relatively small when compared to the front, or upstream, O_2 sensor. Automotive engineers call this storage process "Cat Punch-Thru," and depending upon whether voltage is high or low, the process is specifically called either "Lean Punch-Thru," or "Rich Punch-Thru."

As a vehicle is driven mile after mile, its catalytic converter's performance will ultimately degrade, thus correspondingly causing the converter's ability to store oxygen to also be reduced. In order for an OBD-II system to test a catalytic converter, both upstream and downstream O_2 sensors are used. A worn-out cat in poor operating condition will cause a downstream O_2 sensor's voltage levels to more closely match those voltage readings from an upstream oxygen sensor. When both of the signals from the upstream and downstream sensors reach a pivotal point of equality that is predetermined, a PCM will then set a DTC indicating the catalysts' monitor has failed, and the catalytic converter is no longer operating as it should.

Short History of Fuel Injection Systems

With a final gasp of breath, the use of carburetors on vehicles sold in the United States died in about 1994, with the last of these relics produced by Isuzu. How these mechanical fuel delivery devices came to be replaced by modern electronics will provide some insight and perspective into modern fuel injection systems. In fact, the combination of electronic fuel injection systems with OBD-II diagnostic systems that are equipped with oxygen sensor operation is the only method of fuel delivery that effectively works with catalytic converters. Now that we have covered how catalytic converters and oxygen sensors operate, it's time to tie everything together with a discussion of how electronic fuel injection operates. We'll start out with a brief look back at the origins of automotive fuel injection and follow up with an interview with a very special guest.

Driven by ever-tightening federal emissions standards, the widespread use of electronics for controlling fuel delivery on cars and trucks began in earnest in the early 1980s. The trend toward using solid state electronics in the automotive field started in 1975, beginning with the replacement of mechanical ignition systems by electronic ignition modules, and later by electronic fuel injection systems. However, mechanical fuel injection is not a new innovation in vehicle applications. It was used in diesel engines as early as the mid-1920s, while the first practical use of fuel-injected, gasoline-powered engines dates to World War II, when it was first used on military aircraft. For example, the fuel-injected Wright R-3350, an 18-cylinder radial engine, was used in the B-29 Superfortress bomber.

One of the first commercial gasoline-powered injection systems for automobiles was developed by Bosch and used on the 1955 Mercedes-Benz 300SL. In 1957, Chevrolet introduced a mechanical fuel injection option, made by General Motors' Rochester division, for its 283 V-8 engine. The Chevrolet fuel injection system used a crude method to measure airflow into the engine via a spoon-shaped plunger, which moved in proportional response to the amount of air entering the engine. The plunger was connected to a fuel metering system that mechanically injected fuel into the cylinders. Subsequently, during the 1960s, the popular Hilborn mechanical fuel injection system was widely used in all types of auto racing, but these systems were designed to yield maximum horsepower, and consequently, had relatively bad manners when it came to the more ubiquitous part-throttle driving conditions typically found on America's expanding road networks and developing freeways. Consequently, Hilborn mechanical fuel injection systems were not typically in use on street-driven cars, except for perhaps the occasional hot rod.

One of the earliest electronic fuel injection systems for mainstream automotive commercial use was developed by Bendix; called the "Electrojector," this fuel injection system was used on the DeSoto Adventure as early as 1958. This system was most likely the first electronic throttle body injection (TBI) system ever used in a production automobile. These early fuel injection systems, whether mechanical or electronic, never gained popularity because they were difficult to manufacture, hard to maintain, and did not provide any real benefits to consumers. Consequently, American auto manufacturers went with what they knew for providing fuel delivery in vehicles, and what they knew best was the good old carburetor.

At the time, carburetors were more reliable and less costly to manufacture than any type of fuel injection; however, this was not the case in Europe, where Bosch continued to develop and refine its fuel injection technology. One of the first fuel injection systems developed by Bosch for production vehicle use debuted on a 1967 Volkswagen. Called "D-Jetronic," (the "D" stood for the word "Druck," which is a German word for "pressure"), this system was a speed-density type of fuel injection system. Speed-density simply meant the engine's speed and the density of the air inside the intake manifold were both used to determine the amount of fuel to be injected into the engine. An electro-mechanical pressure sensor (an early form of a modern-day MAP sensor (manifold absolute pressure sensor) was used to calculate engine load, or how much fuel an engine would require for any given operating condition. The D-Jetronic system found its way into use on Volkswagen,

Mercedes-Benz, Porsche, Saab, and Volvo vehicles. General Motors used a copy of the D-Jetronic system on its Cadillac models starting in 1977.

However, the D-Jetronic system was already being replaced in 1974 by the improved L-Jetronic and K-Jetronic fuel injection systems on European cars. The L-Jetronic system ("L" stands for "Luft," which means "air" in German) used a mass airflow sensor (MAF) equipped with a mechanical flapper door that moved whenever an engine's throttle was opened. The door was connected to a variable resistor that sent a voltage signal to an onboard computer. This system used electronic injectors. Often called "port fuel injection," this system used one injector for each cylinder. This type of electronic fuel injection (EFI) system was also popular on Japanese cars during the 1980s, and remained popular on European vehicles as well.

Today, all vehicles made for the American marketplace are equipped with EFI systems. In addition to cars and light trucks, there are many nonautomotive applications for EFI systems as well, including motorcycles, scooters, all-terrain vehicles (ATVs), and even lawn tractors. Consequently, today, professional automotive technicians, as well as home do-it-yourself mechanics who are lacking a basic understanding of how EFI systems work, are severely limited in their ability to perform basic diagnostics, even with the help of a scan tool connected to an OBD-II system.

Starting with simple, basic concepts and comparisons between older carbureted fuel systems and newer EFI systems should offer some insight as to how more modern systems operate. So what do carburetors and fuel injection have in common, and more importantly, what do they have to do with each other? Older carburetor systems and modern EFI systems share similar basic functions and fuel delivery duties. Consequently, a good understanding of how a simpler carburetor's fuel circuits operate will help make sense of how a more sophisticated EFI system operates. Once a technician grasps the basics of EFI operation, most types of electronic fuel injection are relatively easy to understand.

Carburetors

The fuel delivery functions of a carburetor are not substantially different from similar functions newer EFI systems perform. Therefore, understanding how fuel delivery circuits inside a carburetor operate will make electronic fuel injection concepts and operational functions easier to understand. All gasoline-powered engines need only three basic ingredients to run: (1) the correct amount of fuel for any given rpm and throttle opening; (2) a spark from the ignition coil at the right time; and (3) sufficient cylinder compression to ignite the air/fuel mixture. A carburetor is basically a hunk of aluminum with a bunch of holes drilled into it. A carburetor's job is to match the supply of fuel to

Carburetors have been around for over 100 years. This one from the 1970s performed all the same basic functions as a modern fuel injection system. Although carburetors were simple to repair, fairly reliable, and inexpensive to produce, they unfortunately weren't precise enough to provide reliable fuel delivery with the requisite degree of accuracy required by today's emission standards. *Courtesy Elwoods Auto Exchange*

the amount of air entering the engine. The volume of air is controlled by the driver's right foot whenever the driver operates the gas pedal. As air pressure within the carburetor changes, fuel and air flow through the various holes in the carburetor into the engine.

All of the functions a carburetor performs are duplicated in a modern electronic fuel injection system. Carburetors use several hydraulic fuel circuits that deliver fuel to the engine's intake manifold. Fuel moves through a carburetor's fuel circuits as a result of differences in air pressure at various points in its passageways and openings. A carburetor's fuel-delivery system is made up of at least six basic separate fuel circuits, each with a specific job to perform. Following are the basic fuel circuits found in most carburetors.

Float Circuit—The float circuit regulates fuel into a carburetor and stores it in a float bowl where it is available for use by five other fuel delivery circuits. When the level of fuel drops in the float bowl, a needle valve supplies more fuel to keep the bowl full—similar to a valve in a toilet bowl that keeps water in the bowl when the device is flushed.

Choke Circuit—During cold start-up, the choke circuit adds extra fuel and air to the engine to keep it from stalling, until it reaches normal operating temperature. Early carburetors used a manual choke circuit operated by a cable and choke knob on a dashboard, while later models used an automatic choke controlled by a bimetal spring.

Idle Circuit—The idle mixture circuit provides fuel for the engine at idle speeds; this circuit is controlled by an idle mixture screw that regulates the amount of fuel entering the engine when the carburetor's throttle plate(s) are closed.

Fuel Transfer Circuit—The transfer circuit is made up of a series of small holes located in the carburetor's main passageway that progressively add fuel as the carburetor's throttle plate(s) are first opened. This circuit provides smoothly transitioning fuel delivery from idle to part-throttle operation.

Main Circuit—During steady throttle operation, fuel is delivered from the float bowl through the main circuit and into the engine. The amount of fuel is controlled by the main jet, which is basically a small hole, located in the bottom of the float bowl, that restricts the flow of fuel. The size of the main jet is carefully matched to airflow through the carburetor—too large a hole and fuel economy suffers, too small a hole and engine performance feels sluggish or flat at part-throttle.

Accelerator Pump Circuit—Whenever the throttle is opened suddenly, the accelerator pump circuit squirts fuel directly into the intake manifold. Since air is about 400 times lighter than gasoline, air would normally reach the intake valve and combustion chamber well ahead of the fuel delivered by the carburetor. Thus, this "squirt" of fuel from the accelerator pump circuit allows the main fuel circuit to play "catch-up" with the column of moving air entering the engine. The accelerator pump circuit prevents a flat spot and possible backfire during abrupt engine acceleration.

Although carburetors have been performing fuel delivery functions in vehicles for well over 100 years, and have always basically worked pretty well and pretty much the same, there are some functions carburetors simply can't handle. Unfortunately, a carburetor's ability to deal with constantly changing operating conditions is limited with regard to its reactions to changes in altitude, its efforts to compensate for engine temperature, its lack of precise control of fuel for emissions' purposes, and its overall use of too much fuel during steady-state engine operation and acceleration. In a word, carburetors are just plain too "dumb" to continue being able to provide accurate air/fuel mixtures in modern vehicles. What is needed is a fuel delivery system with brains.

Electronic Fuel Injection

The design and concept behind all EFI systems is essentially the same: A computer, with various engine sensors providing electronic information, is used to control the exact period of

Similar in design to a motorcycle carburetor, this Ford variable venturi (VV) carburetor rivaled some electronic fuel injection systems in both engine performance and fuel economy. VV carbs were used throughout most of the 1980s; many police cruisers used them because of the exceptional power and throttle response they provided. *Courtesy Elwoods Auto Exchange*

This single-barreled, cast-iron carburetor would best be used today as a paperweight, or possibly as a replacement part in a classic car or truck awaiting restoration. Amazingly, there are still vintage auto parts' suppliers online that sell rebuild kits for this old hardware. *Courtesy Elwoods Auto Exchange*

This Ford Motor Company throttle body and four injectors comprise a modern EFI system for a four-cylinder engine. Because each cylinder has a source of fuel located close to its intake valve, there is virtually no lag between throttle opening and fuel delivery. Better overall engine performance and fuel economy give EFI systems many advantages over carbureted systems.

time fuel will be injected into an engine. In addition to performing all of the functions of a carburetor, EFI systems also control engine idle speed and ignition timing system functions. They primarily regulate fuel delivery via electromagnetic valves (fuel injectors) that electronically turn on for precise lengths of time as measured in increments of thousandths of a second, or milliseconds. Once the fuel injectors are turned on, they spray fuel into the engine. The amount of "on" time is called injector pulse width, and the longer this is, the greater the amount of fuel injected. The onboard computer, or PCM in an OBD-II system, is continuously receiving data from its various sensors and calculating how much injector "on" time should be used for fuel delivery.

While all EFI systems use fuel injectors that operate in a similar manner, several different EFI designs and computer strategies existed prior to the inception of OBD-II systems in 1996. One of these designs was throttle body injection, or TBI, which used only one or two injectors located where a carburetor would normally be installed. In this type of system, after fuel is injected into the throttle body, it must travel through the intake manifold before reaching each cylinder. Since the air/fuel mixture has to travel different distances to reach the engine's cylinders, this system had inherent fuel distribution problems, as it caused differing amounts of fuel to reach the cylinders. Thus, consistency in fuel delivery was lacking. Consequently, TBI injection is no longer used in view of its inherent inability to regulate equal amounts of fuel to each cylinder, and in the process, effectively control emissions.

Unlike TBI injection, port fuel injection (PFI) uses an individual injector for each cylinder. A PFI system is more efficient than a TBI system, because fuel is sprayed directly at the back of the engine's intake valves; consequently, it has less distance to travel before getting to the combustion chamber, rendering fuel delivery more precise. However, PFI systems were eventually phased out of production and replaced by more sophisticated sequential port fuel injection (SPFI) systems. The injectors on SPFI systems are pulsed in accordance with an engine's sequential firing order—similar to how ignition systems are timed and spark plugs are fired. Thus, SPFI systems provide even more accurate fuel delivery, which in turn, helps reduce exhaust emissions and increases fuel mileage and performance.

This throttle body electronic fuel injection unit is from a Ford EEC-III vehicle from the mid-1980s. Two Bosch high-pressure injectors are used (at the top with wires coming out of them). The fuel pressure regulator can be seen on the top at left. These types of systems provided better drivability than a simple carburetor, got fairly good fuel economy, and were able to meet emissions requirements of their day. *Courtesy Elwoods Auto Exchange*

More sophisticated port fuel injection (PFI) systems provided more accurate fuel delivery than throttle body injection systems since these systems provided individual cylinders with their own source of fuel—a fuel injector. On this Mitsubishi PFI system, the individual injectors are sandwiched between the cylinder head and fuel rail. In addition to separate electronic fuel injectors, each cylinder also has an individual intake runner. The lengths of the intake runners allow more air to enter the engine at higher rpm, thus boosting the amount of torque the engine produces. *Courtesy Elwoods Auto Exchange*

Electronic fuel injectors like this General Motors example are nothing more than simple electro-mechanical valves that operate via electronic pulses from a vehicle's PCM. Power is continuously supplied to the injector, while the PCM pulses the ground return wire to turn the injector on and off. Fuel injectors are fairly simple to test, as they only require a digital ohmmeter to test injector resistance, and a test light or noid light to test for injector pulse from the PCM. Unfortunately, getting to an injector may be tough, depending on how buried they are under engine covers and intake manifolds. *Courtesy Elwoods Auto Exchange*

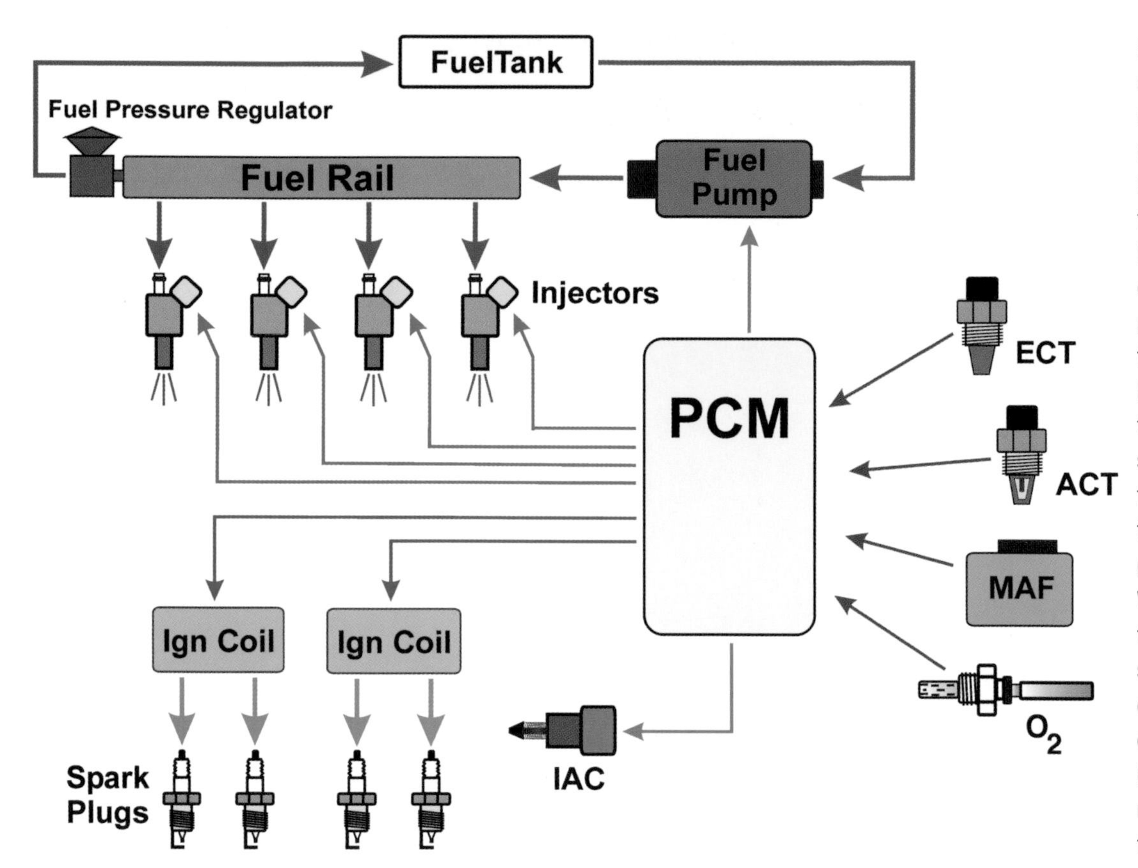

Fig 3-7—Illustrated is a basic engine management control (EMC) system. The PCM controls both the fuel delivery and ignition systems. An electronic fuel pump pressurizes fuel in the fuel rail, which acts like a storage tank for the injectors. In the system shown, excess fuel is returned to the fuel tank via the fuel pressure regulator. When the PCM pulses the injectors, fuel is sprayed into the engine. Based on computer sensor inputs (shown at far right), the PCM also fires both ignition coils as the engine rotates.

Finally, one other type of fuel injection system exists: Bosch's "K" or "KE-Jetronic." This system uses an airflow sensor in which a metal plate moves in relation to engine airflow. As the throttle is opened, the sensor measures the increased amount of air entering the engine. A fuel distributor then delivers fuel to each injector mechanically. These systems are found only on European cars like Mercedes, BMW, Volkswagen, Saab, and BMW. The last "KE-Jetronic" system manufactured for the U.S. market was equipped on a 1993 VW Passat.

Hopefully, readers now have a better idea about how electronic fuel injection, oxygen sensors, and catalytic converters operate. With a little history under the belt and some solid working knowledge of the operations and relationships of various vehicle systems, it's now finally time to move onto the subject of scan tools and code readers. Since scanner data streams and related diagnostic trouble codes are so closely interwoven with engine management systems, having a basic idea of how the fuel/air chemistry operates in an internal combustion engine will help with understanding the information scanner tools provide, as well with diagnosing and interpreting OBD-II system diagnostic trouble codes and their root causes.

This K-Jetronic fuel injection system is mostly mechanical in operation. Bosch developed this form of fuel injection as a result of their experience with diesel engine injection systems. These systems were reliable and fairly easy to repair, but it was expensive to do so. They could not control fuel delivery precisely enough to be used on any OBD-II systems, and were phased out of production in 1993. Prior to 1993, they were used on Volkswagen, Saab, Mercedes, Volvo, and BMW vehicles. *Courtesy Elwoods Auto Exchange*

INTERVIEW WITH A PCM

Fig 3-8—Mr. PCM here has the computing power to manage his vehicle's fuel and ignition system. But is he smart enough to deal with the intricacies of an OBD-II diagnostic system and all it entails? Find out as you read the first interview with a PCM ever recorded. *Courtesy Alan Lapp, Level Five Graphics*

While a carburetor may undoubtedly be a dumb hunk of aluminum, there are those who would argue a PCM (power control module) is nothing more than a dumb wafer of silicone with wires attached. To settle the matter, we searched for and found a PCM willing to be interviewed, and ready to answer questions about how electronic fuel injection systems work. This OBD-II-compliant PCM has more smarts than most, so we're lucky to have this opportunity. In this Motorbooks Workshop series exclusive, the first ever "face-to-processor" interview, we got the chance to ask all the questions about EFI systems that everyone's been dying to have answered.

Motorbooks—Thanks, PCM, for agreeing to talk with us. We don't often get a chance to interview automotive onboard computers. Can you introduce yourself and give us a general description of what you do?

PCM—Thanks for having me! It's so great to hear that people are interested in what I do. Some less enlightened people have unkindly referred to me as an overgrown calculator. While there is some truth to that description, it's not exactly true, since I do have more computing power than a calculator, though less than a typical laptop. I have really fast access to my memory and can perform close to a million calculations per second. Plus, I have a network of friends that supply me with information when needed. You see, I have connections—a bunch of sensors that feed me information on various engine-operating conditions. These sensors serve me at a moment's notice, acting like my own eyes and ears. It's discerning friends like these that allow me to keep things under control. I can also calculate fuel delivery and ignition settings that correctly respond to whatever the "nut" behind the steering wheel might have in mind while driving the vehicle I'm installed in.

MB—Hmm. That is impressive. Sounds like you don't often get enough credit. Why don't you take us through the steps you would typically go through while managing the engine and its various systems? For instance, what do you do when the ignition key is first turned to the "On" position?

PCM—I take voltage readings from the engine coolant temperature sensor as well as the outside air temperature sensor. To illustrate, assuming the outside air and engine temperatures read 50 degrees Fahrenheit, I'll increase the injector "On" time,

While they might look different, all of these fuel injectors operate in the same manner. Each uses a coil of wire that acts like an electromagnet to open a valve whenever the PCM sends out an injector pulse triggering fuel delivery to the engine. Injectors are either all the way open and on, or completely closed and off. A PCM takes readings from all of its sensors and calculates the injector on time, or pulse width. How long the injectors are on and open determines how much fuel will be injected into the engine.

This General Motors MAP sensor serves as the equivalent of an electronic vacuum gauge that measures engine vacuum. Readings from this sensor are used by a PCM to calculate fuel injector pulse width based on engine load, or how hard the engine is working. All MAP sensors provide the same information, but they send out different electrical signals depending on the specific manufacturer. General Motors and Chrysler MAP sensors produce a varying voltage signal, while Ford MAP sensors produce a digital on-off signal. Furthermore, each MAP sensor may be connected differently to an intake manifold, since some auto manufacturers use a vacuum hose, while others mount the MAP sensor directly onto the intake manifold. *Courtesy Elwoods Auto Exchange*

or pulse width, to add extra fuel to start the engine. But I also command the idle air controller, or IAC, to provide some extra air to the engine to increase its idle speed when it starts, otherwise the stupid thing would stall. I'll also check in with my barometric pressure sensor, or BARO, to determine what altitude I'm operating at, so I can fine-tune and match fuel delivery to atmospheric pressure or altitude. Oh, I almost forgot, I also have to turn on the fuel pump for a couple of seconds to prime the fuel system. I can't leave it running for safety reasons, unless of course, the engine is already started and stays running.

MB—Wow. That's a lot. I had no idea you were so busy when the engine starts up. I guess I just took engine starting for granted. Once the engine is running, is there anything else you do to control basic engine operation?

PCM—Yes. In fact, it often feels like my job is never done. Basically, to manage engine operation I must control four things: (1) injector "On" time; (2) ignition timing; (3) engine idle speed; and (4) fuel pump operation. However, please understand that these four functions are the bare minimum I must perform in order to keep the engine running smoothly. To be truly efficient, I must also perform numerous additional functions, like controlling the cooling fan to keep the engine from overheating and shifting the automatic transmission. At the same time, I am often called upon by the driver to control the heating and air conditioning or the lighting system. Furthermore, to show the driver I really care about safety, I also use a "drive-by-wire" system as a precaution; this system controls the throttle so the tires won't lose traction when the road is wet or slippery. In this way, I can prevent certain irresponsible drivers from doing burn-outs and donuts. Finally, I also control lots of other, littler, more incidental stuff that I need not bother you with now. And not to brag, but I should point out that not all PCMs have as many sensor inputs as I do, nor do they control as many vehicle systems as I do. I guess I'm just a little bit smarter than most PCMs.

MB—Truly, I had no idea your job was so demanding and stressful. I admire you for keeping things together so well. Recently, I've been hearing about this thing called an OBD-II system, or onboard diagnostics II. What can you tell me about it? Are you required by the EPA to use OBD-II?

PCM—Yeah, I've not only heard about OBD-II, I have to deal with one in my vehicle—kind of like a bad case of the flu that never goes away. I am required by the EPA to work with one, but I keep the OBD-II thing locked deep in the recesses of my software and occasionally feed it electrons. First, I want to officially go on the record and state that OBD-II has absolutely nothing to do with engine management—that's exclusively my job. The OBD-II system is separate and more like "Big Brother." It's always watching me, waiting for something bad to happen. As a computer, I consider myself perfect in every way, always functioning at optimal performance, but on rare occasion, like anyone else, if one of my sensors or actuators or part of the ignition system screws up, I have hell to pay. When this happens, the OBD-II system writes it down as a fault code, or DTC, electronically speaking. If things get bad enough, my OBD-II software will turn on a "Check Engine Soon" MIL light to alert the driver that I am functioning under duress, with a specific malfunction occurring somewhere in one of my systems. To sum up, I have a close professional working relationship with my OBD-II system counterpart, but we're not particularly close otherwise.

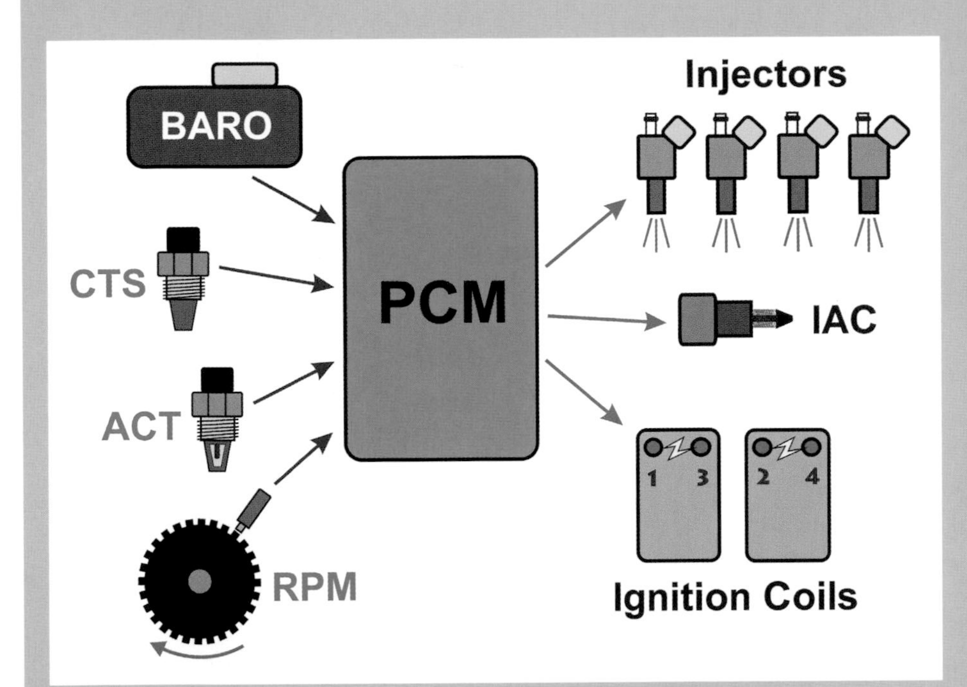

Fig 3-9—Whenever an engine's starter motor is cranked over, the PCM receives an rpm signal from the crank sensor, a barometric pressure signal from the BARO sensor, and temperature readings from the coolant temperature and air temperature sensors. Based on this compiled information, the PCM pulses the injectors and fires the ignition coils, which in turn fire the spark plugs. When starting conditions are cold, the PCM will increase engine idle speed using an idle air controller (IAC) in order to prevent the engine from stalling.

MB—Gee, I had no idea. Sounds tough, but it sounds like you've found a way to work it all out smoothly. Let's get back to the reason for this interview—namely, how electronic fuel injection systems operate. What happens when the ignition key is turned from the "On" position to the "Start" position and the engine cranks over?

PCM—As soon as the engine starts rotating, I receive pulsing signals from both the crank and camshaft position sensors. These signals let me know the engine is trying to start; so the first thing I do is turn the fuel pump back on. Since I know the engine is cold, I'll send signals to all the injectors to spray fuel each time the engine makes a complete revolution. This adds extra fuel required for cold starting conditions. Once the engine is warmed up, I'll use a camshaft position signal to determine which cylinder is about to open its intake valve. Then I'll pulse, or energize, individual injectors in the same firing order as the spark plugs in order to create an atomized cloud of fuel that will wait right at the intake valve, so that as soon as the valve opens, the cloud will get sucked into the cylinder. This strategy is called "sequential port fuel injection" because the injector firings are timed with each cylinder's intake valve opening. Older EFI systems that aren't as sophisticated as I am can only pulse all of the injectors simultaneously every other crankshaft revolution. That's why those older systems had a hard time controlling emissions and didn't generate as much engine power as my system. That's why they're not around anymore, but I still am.

MB—Well, I can see why you would be. That's a lot of work to perform, and it sounds fairly sophisticated. But once the engine is started, is your job done? Can you kick back and relax? Or does something else happen?

PCM—Well that depends. If the driver lets the engine idle for a while, the engine temperature will increase. Then I have to shorten the injector pulse width so that less fuel is delivered to the engine. I'll also gradually lower the idle speed. However, if the driver shifts the transmission into drive and steps on the gas, then I have to jump into high gear to really make things happen quickly, since I am required to add fuel as the engine requires it. I rely on four different informational signal inputs to determine appropriate base injection "on" time, or pulse width, including engine rpm from the crank sensor, camshaft position, throttle position from the TPS sensor (sometimes referred to as driver demand), and engine load from the mass airflow (MAF) sensor. These input signals allow me to adjust and control injector pulse width and ignition timing to match exactly the needs of the engine at any moment.

MB—Wow. Don't you get stimuli overload? With all the information coming in from your sensors, how do you actually decide what action to take in response to either driver demand or engine power requirements?

PCM—Sometimes it's difficult, but I can handle it since I have help. I use three-dimensional mapping programs, sort of the equivalent of visual aids, that are burned into my read-only memory, or ROM. As engine rpm increases, I simply fire the injectors sequentially so that their firing matches engine speed. At any given rpm, I can monitor the MAF sensor to determine the quantity of air entering the engine. Airflow is directly proportional to engine load (how hard the engine has to work to get the vehicle moving down the road). My fuel and ignition maps contain what are known as "look-up tables," where predetermined fuel delivery values are stored. From the information gleaned from the "look-up tables," I can calculate how long it is necessary to keep the injectors turned on, and when to fire the spark plugs. I have to look up these values each time I pulse an injector or fire a spark plug, so I have to think and react pretty quickly. As you can imagine, things happen pretty fast on a high-revving engine.

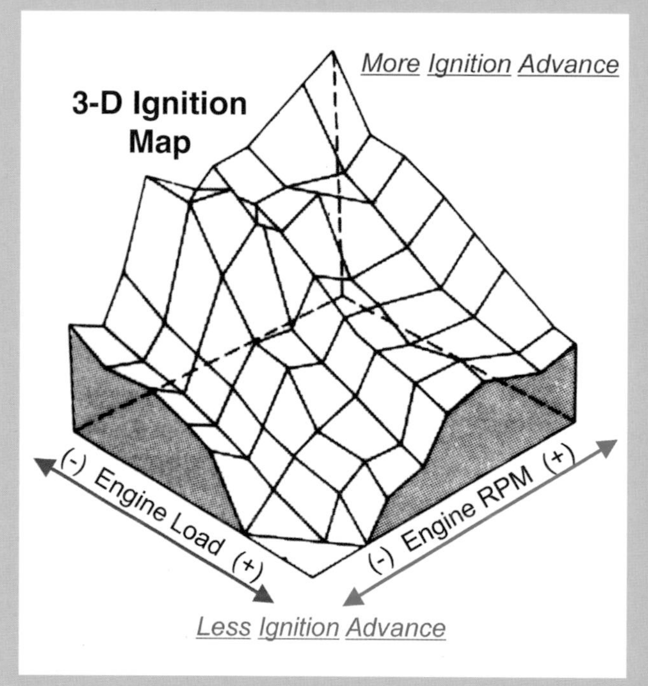

Fig 3-10—This illustration is of a three-dimensional (3-D) ignition advance map, which depicts the relationship between engine load, ignition advance, and engine rpm. While an ignition map is shown, fuel maps are similar in that they are nothing more than a simple set of instructions the PCM follows in order to calculate fuel delivery or ignition timing commands. In this example, less ignition spark advance is used at low engine rpm and high engine loads to prevent engine knock or detonation. As engine load decreases and engine rpm increases, more spark advance is used to increase fuel economy. Because hundreds of engine load and rpm combinations are possible, it takes many hours of engine testing on a dynamometer to produce both 3-D ignition maps and fuel maps.

MB—Life in the fast lane, huh? But I'm still a little puzzled about something you said earlier, when you talked about driver demand and input from the TPS sensor. Can you explain the relationship between these things?

PCM—Sure. The TPS or throttle position sensor is really the only input signal a driver has direct control over. Its varying voltage signal tells me what the driver is trying to accomplish in the horsepower and torque department. No matter how slow or fast the accelerator pedal is pressed to the floor, I can react faster than the driver to provide proper ignition timing and fuel delivery. In fact, the TPS signal is much like an accelerator pump on a carburetor system, as it gives the engine an extra shot of fuel to match increased airflow when the throttle is opened. Let me provide you with a different perspective on this issue: Whenever the throttle is opened abruptly, a column of air moves through the intake manifold at high speeds. To prevent the occurrence of a "flat spot" during acceleration, gasoline must join with moving air before the intake valve opens. To ensure this occurs timely and properly, I can modify the length of time the injectors stay on, at a computer processing rate of over 100 times per second, so that when the throttle is whacked hard I have the exact amount of fuel waiting to mix with the blast of incoming air. Keep in mind that electricity still travels faster than any action either the driver or engine can perform, so keeping up with changing engine speeds and engine loads is just not a problem for me.

MB—Well, that's understandable. Now that I know what a TPS does, I'm curious about something that used to happen quite often with older carbureted systems. When driving in the mountains, engines would run poorly and carburetors would have to be adjusted for higher altitude conditions. Do you experience a similar phenomenon, and if so, how do you deal with this problem?

PCM—I am happy to say that scenario is a nonissue for me. I simply have to read the input from my barometric pressure sensor, or BARO sensor, and then just adjust the fuel delivery to match the altitude the car is being operated at. The higher

This Bosch MAF sensor uses a hot wire system to measure airflow into the engine. The hot wire, which is located inside the small ring or tube, is visible and is heated by the sensor's electronics. When the throttle opens, incoming air blows past the heated wire, cooling it down. An MAF sensor's internal microprocessor increases electrical current to the hot wire to maintain the wire's temperature at a predetermined point. By sensing the amount of current required to keep the hot wire at the proper temperature and comparing it with a measurement of the outside ambient air temperature taken by an additional temperature sensor, the MAF can then calculate engine airflow and even determine relative outside humidity. MAF sensors are used more often today than MAP sensors, since they are far more accurate for determining engine airflow.

the altitude the car is driven, the less air or oxygen is available for combustion. Consequently, I simply shorten the pulse time, or pulse width, of the fuel injectors and add less fuel to the engine. The incremental adjustment I make is called fuel trim, and it provides the engine with exactly the right amount of fuel, regardless of the altitude the vehicle is operated at. Plus, I also use other sensors to trim or fine-tune fuel delivery. As an example, when the air charge temperature (ACT) sensor notifies me the vehicle is operating on a hot day, I adjust optimal fuel delivery parameters and remove a small amount of fuel. On a cold day, I simply do the reverse—I add a little more. I am also able to calculate relative humidity using information from the MAF sensor, and can adjust fuel delivery to accommodate associated changes in air density as well.

MB—While all this electronic fuel control wizardry is occurring, how do you determine proper ignition timing, or when to fire the spark plugs?

PCM—Again, I rely on information from my assistants, or sensors. Depending on specific informational inputs I receive from a TPS sensor, engine speed or rpm, and MAF sensor, I either advance or retard ignition spark timing depending on engine demand. For example, if the driver is operating the engine at part throttle combined with moderate engine rpm, I will adjust the ignition timing by advancing it, or starting it earlier for better fuel economy. However, if the reverse occurs and the throttle is opened suddenly, and engine rpm are low, then I must adjust the ignition timing accordingly, by retarding it, or starting it later to prevent engine knock; then later, as rpm increase, I will again advance ignition timing to always ensure optimum horsepower production.

MB—You mentioned engine knock. I'm not all that familiar with that. Isn't that a funny noise a vehicle makes? Is there such a thing as a knock sensor? If there is, and I am betting there probably is, how does it work?

PCM—You would win that bet. There is a knock sensor, and its sole purpose is to detect whenever the engine knocks, pings, or detonates. All of these conditions are bad and can cause severe engine damage. Once I receive a signal from the knock sensor, I slightly retard ignition timing until the knocking stops. In fact, I am smart enough to tell which cylinder is knocking, so I will only retard the spark timing for that specific cylinder. Once the ping or knock disappears, I'll immediately advance ignition timing again for better fuel economy and engine power.

MB—You haven't talked about oxygen sensors or catalytic converters yet. What's going on with these emissions devices, and what's your role in their operation?

PCM—Well, a catalytic converter, or "cat" as many people commonly call it, controls engine exhaust emissions that occur during two engine-operating conditions—idle and part-throttle. For the cat to work properly, the fuel injection system must be able to average its air/fuel ratio to just slightly above and slightly below 14.7:1, the ideal proportional relationship that's 14.7 parts of air to 1 part of fuel. To accomplish this, I use my upstream oxygen sensor, which is located between the engine and the cat. This sensor measures oxygen content in exhaust gas and as a result of these measurements produces a small signal voltage that it sends to me. I then determine what fuel corrections to make to keep the air/fuel ratio averaging 14.7:1. This scenario goes something like this; I inject fuel into the

This three-wire General Motors MAF sensor produces a high frequency signal that is sent to a PCM, which uses the informational signal to calculate total airflow into the engine. Airflow calculations combined with intake air temperature factor in the volume of air entering the engine with altitude and humidity determinations as well. Less sophisticated MAP sensors were eventually replaced by MAF sensors on many vehicles, since they provide faster, more accurate engine load information to a PCM. *Courtesy Elwoods Auto Exchange*

engine, *sense* exhaust gas oxygen and *compare* it to what is needed for an ideal mixture and *correct* subsequent fuel delivery. This strategy is called "closed-loop" operation because it is an informational loop—sense, compare, and correct—get it? I should add that closed loop only takes place when the engine is hot and either at idle or part-throttle.

MB—OK, if that's how closed-loop operation works, is open loop operation just the opposite?

PCM—Let's see. How can I explain this? OK, remember how I only use closed-loop operation during idle or part-throttle driving conditions? Well, I use open-loop operation for all other driving conditions. For example, cold or hot starting, accelerating, decelerating. All of these modes of engine operation do not require an ideal 14.7:1 air/fuel ratio that I spoke of earlier. So the "sense, compare, and correct" form of informational loop is not used or necessary, hence the system is now operating in open-loop mode.

MB—Well, that's certainly a lot to think about. You sure do know your stuff. It's great that you took time out of your busy schedule to speak with us. Thanks for answering our questions. It was great talking with you.

PCM—My pleasure. Thank you for having me. Interacting with humans can be mildly interesting and sometimes amusing.

MB—I guess that wraps it up for now. I'd like to thank my audience, too, for being part of an exclusive interview with a power control module about how it manages a vehicle's fuel and ignition systems. This is Motorbooks International signing off. Have a great day.

CHAPTER 4
PROFESSIONAL SCANNERS AND CODE READERS

This OTC Monitor 2000 was one of the first aftermarket scan tools available for independent repair technicians. It initially only worked with older, first-generation OBD General Motors vehicles, but as other manufacturers improved and updated their OBD-I systems, the Monitor 2000 could be used to read codes and data on vehicles from Ford Motor Company and Chrysler as well. *Courtesy OTC Tools*

Introduction

Automotive scan tools have been around since the early days of OBD-I diagnostic systems. In the 1980s, these electronic scan tools were limited in what functions they could perform, as their capabilities often depended largely on which manufacturers' vehicle they were plugged into. New car dealerships had factory scanners that could access a greater degree of information (some of it proprietary) than aftermarket scan tools could, but manufacturers' scan tools were simply not available to independent repair shops. Back then, all aftermarket scanners were expensive, and as such, were typically beyond the economic reach of the average backyard auto mechanic. Both original equipment manufacturers' (OEM) scan tools and aftermarket scan tools improved substantially over the next dozen years or so in the diagnostic functions they could perform, along with the OBD-I systems they were used in conjunction with on cars and light trucks.

By the mid-1980s, early do-it-yourself code readers started showing up in Sears, Pep Boys, and other auto parts chain stores. These early code readers were capable of reading diagnostic trouble codes from the computers installed in vehicles manufactured by General Motors, Ford Motor Company, Toyota, and Honda. The earliest code readers used LED lights to flash out trouble codes, and were similar to a check-engine light (MIL) on some vehicles, since they were not capable of providing any other information. These simple code readers provided home technicians with an easy way to read trouble codes without needing to use a test light or jumper wires, and most importantly, without needing to follow complicated procedures for trouble code retrieval.

Today, modern code readers and scan tools designed for use with OBD-II systems, and specifically for the do-it-yourself market, have a wide variety of capabilities. Most models of scanners and code readers are priced from less than $100 to more than $700, and are therefore more affordable for the average automotive consumer. Once any

This late 1980s model OTC Monitor 4000E scan tool was one of the cutting-edge professional scanners available at the time for the aftermarket industry. Data cartridges pertaining to specific brands of vehicles could be inserted at the base of the unit to provide greater manufacturer-specific information. In addition, an OTC Pathfinder software cartridge could also be used to display generic tune-up specifications and other useful diagnostic tips and information. This OTC Monitor 4000E scanner was one of the first aftermarket scan tools with OBD-II capabilities. *Courtesy OTC Tool Company*

of these newer scan tools, or code readers, are connected to an OBD-II system, they are capable of providing more useful diagnostic information than the majority of professional scan tools of the past. In fact, some of the scan tools available for the home market today rival the capabilities of professional tools of yesteryear, but for a fraction of the price. In this chapter, we'll take a closer look at the development of professional scan tools and code readers, and will focus how to use, and the many uses of, code readers. Chapter 5 will cover the use of scan tools in greater depth, as well as PC/laptop interface scanners, and personal digital assistant (PDA) scan tools.

Professional Scan Tools

Professional scan tools are of two basic types: OEM dealership-specific and generic aftermarket. Factory- or manufacturer-specific scan tools are only available to new car dealerships for exclusive use on specific makes of vehicles they sell and service. Dealership scanners have all the latest, greatest information available directly from OEMs about the vehicles they manufacture, including technical service bulletins (TSBs), factory recalls, and other proprietary information. Some of these scanners provide a direct link to the factory, where scanned information about new vehicles is stored in a central database. Factory scan tools also allow professional technicians to perform many more diagnostic tests on specific system components. Most factory scan tools are wireless and can therefore scan cars or trucks in a service area, on the showroom floor, or out on a lot, simply by entering a specific vehicle's VIN number. Oftentimes, the resulting fix for drivability issues or emissions-related problems on new cars or trucks can come down to a simple need for revisions to a vehicle's OBD-II operating system. OEM scanners can also be used to reprogram a vehicle's PCM with the latest factory software updates.

Since professional-grade aftermarket scan tools offer more functions and more current, comprehensive information than do-it-yourself scanners, they consequently cost more to own and update. These scan tools typically cost from just more than $1,000 to more than $6,000, depending on options and accessories equipped. The old adage that "you get what you pay for" definitely applies to these tools,

This Toyota handheld scanner is now retired and lives in a storage cabinet in the service area at the dealership. However, this diagnostic tool is not yet obsolete, as it is still used as a backup whenever the newer scan tool has a problem. This device is still capable of providing a professional technician working on Toyota vehicles with more vehicle-specific information than most generic aftermarket professional scan tools. *Courtesy Younger Toyota*

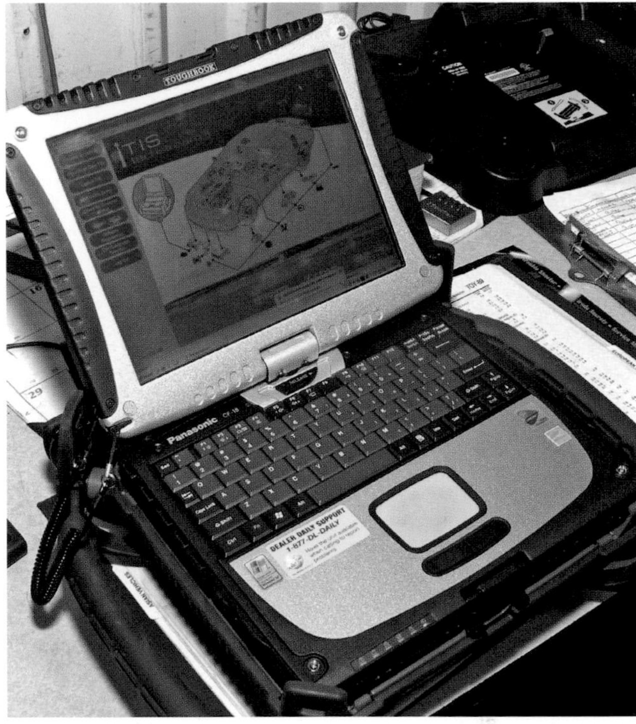

As of publication, the photograph shows the latest Toyota/Lexus/Scion scan tool, which is basically a combat-ready laptop computer built to military specifications that has been "ruggedized" (military jargon) for military and other field use on vehicles. This diagnostic computer system is wireless and can be directly connected to any vehicle for purposes of performing test drives. The software accompanying the diagnostic program is Microsoft Windows-based and easy to use, with data stored and displayed in numerous ways. Software updates are easily accomplished, simply by turning a vehicle's ignition key to the RUN position and clicking on the scanner's mouse pad to "Update." Data transmission is wireless, so the vehicle being scanned can be located anywhere within a dealership—the service bays, vehicle storage lot, or showroom. *Courtesy Younger Toyota*

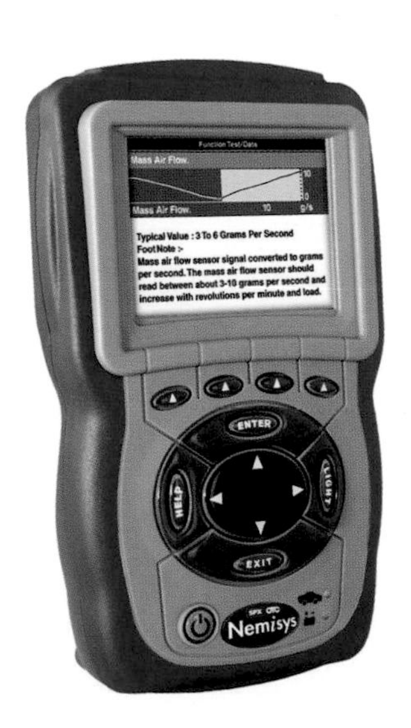

The Genisys System 2.0 scan tool shown offers a wealth of diagnostic information to professional technicians. Automakers, including General Motors, Ford Motor Company, Chrysler, Toyota, Honda, BMW, Volkswagen, and others, maintain licensing agreements with SPX/OTC Corporation (the manufacturers of the Genisys System) to continually provide the latest vehicle-specific data for use with the Genisys scanner. This Genisys professional scanner is available with Pathfinder software that provides DTC information, data/sensor information, PCM connector information, drivability symptoms, TSB references, a component locator, and vehicle specifications. Even older, pre-OBD-II vehicle information is updated on a yearly basis. Software updates are provided free via the Internet, so this scanner's reference information is easily maintained. *Courtesy SPX/OTC Corporation*

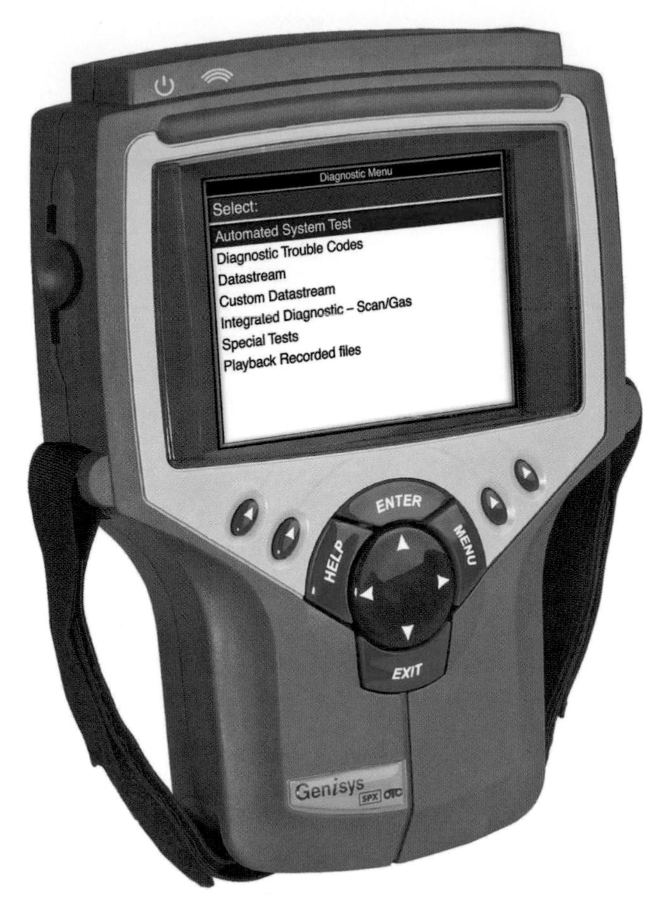

Just like a Genisys scanner, this OTC Nemisys scan tool can interface with both OBD-I and OBD-II vehicles. This tool comes complete with ScanMate software that allows it to upload and download information to a PC or laptop. Both domestic and Asian-enhanced OBD-II information is accessible, as well as repair information that lists typical sensor and data specifications. *Courtesy SPX/OTC Corporation*

as they are truly designed with a professional technician in mind. Since another old adage—time is money—is also true, in the hands of an educated professional automotive technician, these scan tools are capable of quickly diagnosing complex drivability problems since they operate at faster processor rates, have more memory than less expensive scan tools, use enhanced diagnostic features, and offer an abundance of vehicle-specific information that is typically provided directly from automakers. In addition, many of these tools can also function as lab scopes (oscilloscopes), displaying waveforms from computer sensors and actuators.

Many professional scanners perform automated tests that provide an instant snapshot of vehicle health. Specific information related to DTCs, along with factory technical service bulletins (TSBs) and specifications for repairs and replacement parts make these tools more useful and desirable than an OBD-I or -II generic aftermarket scanner that simply reads and erases diagnostic trouble codes. Other information that professional scanners typically have access to includes detailed DTC information, a database of common "drivability" problems (as well as their potential or recommended fixes), a component locator, data/sensor information, and PCM connector information. In addition to providing full coverage of engine management diagnostic issues, other systems are also covered, including ABS braking systems, electronic steering and traction control, heating and air conditioning (HVAC) systems, computer-controlled charging systems, electronic suspension, supplemental restraint systems (air bags), and CAN-controlled electrical systems. Even OBD-II cables can be configured for specific vehicles by using "smart" inserts that allow enhanced manufacturer-specific component testing.

Since professional scan tools often use software that is specific to a vehicle's year, make, model, and engine, bidirectional testing of various systems and individual components can be performed. This means that the scanner can not only receive information but can send out commands to the OBD-II system as well. With bidirectional testing, a user can program a scan tool to command various components to operate. This allows a quick way to verify that the PCM is in control and can command a solenoid to operate, ignition coil to fire a spark plug, or fuel injector to pulse fuel into the engine. Professional scanners can be connected to a PC or laptop to store scanned information, or to upload or download information via the Internet. Software updates from scanner manufacturers are often available on a quarterly basis; thus, professional scan tools are typically kept more current than less expensive scan tools. However, due to the high cost of professional scanners, most home technicians would have a tough time justifying their expense. But not to worry—a professional grade scanner is not necessarily the only effective tool for performing diagnostic work on OBD-II system vehicles. Fortunately, do-it-yourself electronic diagnostic tools can also provide a great deal of OBD-II-related information. Let's take a more detailed look at the most basic OBD-II system interface—code readers.

Code readers are the most basic of OBD-II electronic interfaces. Their primary function is to simply read and erase diagnostic trouble codes (DTCs). The majority of all code readers (and all of the ones covered in this chapter) provide information about which OBD-II inspection and maintenance (I/M) monitors have been run, or are listed as "completed." I/M monitors are tests that the OBD-II system runs to verify that all its systems and components are

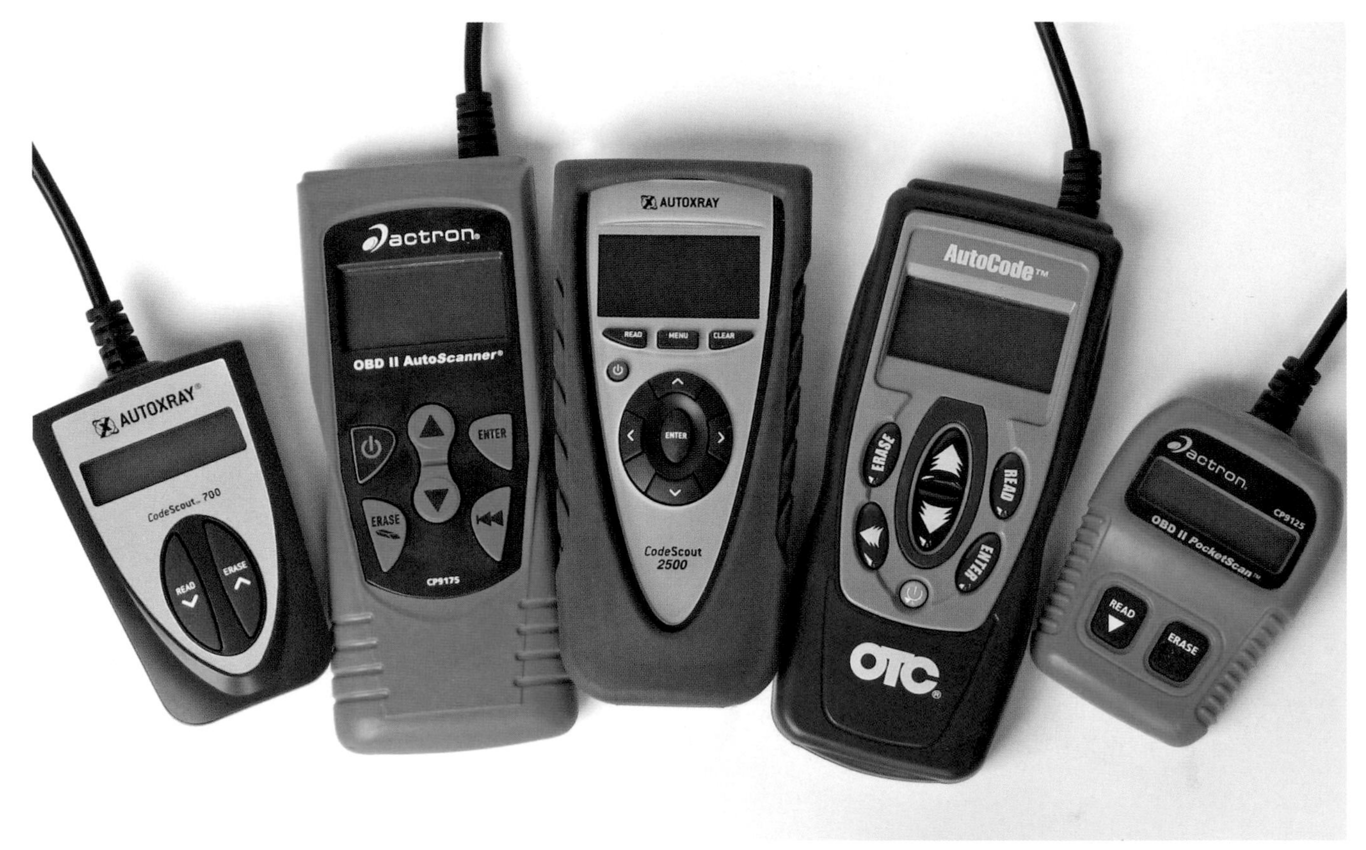

The five code readers shown in the photo range in price from less than $100 to just under $250. They provide the basics for working with OBD-II engine management systems found on 1996 and newer cars and light trucks. Most basic code readers have standard features that include diagnostic trouble code reading, clearing trouble codes, and a self-inspection/maintenance (I/M) readiness monitor status. *Courtesy Actron, Autoxray and OTC Tools*

operating as they should. (See Chapter 2 for more information on I/M monitor status.) Code readers are also capable of indicating the status of whether a malfunction indicator lamp (MIL) is on or off. Some advanced model code readers come equipped with their own built-in reference library of internal diagnostic trouble code definitions, and can display OBD-II freeze-frame data as well. Software can also be easily updated on some models when they are connected to a personal computer with Internet access.

Despite their no-frills image, all code readers are ideally suited for determining if a vehicle is ready and able to pass in-state emissions' testing. They are also useful for verifying the need for recommended automotive repairs whenever a car or truck is taken to a dealer for servicing, since any DTCs set by a PCM can be read by a code reader and written down prior to dropping the vehicle off, and then that list is compared with whatever information the service writer says is wrong with the vehicle.

However, code readers are limited in their ability to perform engine management diagnostics, since they are primarily designed as a general informational tool, and as such, they will not display OBD-II data streams. Typical code readers are generally priced from around $80 to just under $250, depending upon features. We'll take a closer look at five code readers and their diagnostic capabilities, starting with two of the least expensive, both of which happen to be the easiest to use.

CodeScout and PocketScan Code Readers

The CodeScout model code reader from Autoxray and the OBD-II PocketScan code reader from Actron have similar features and a simplified method of operation. Once either of these code readers is connected to an OBD-II-compliant vehicle, either tool is automatically capable of determining which type of OBD-II system communication protocol is being used by a PCM; both can also read newer controller area network (CAN) data protocol. Furthermore, both tools can read and erase generic DTCs, display MIL status, and display I/M monitor status. In addition, these code readers don't require internal batteries for power, but rather, use a vehicle's data link connector (DLC) as a 12-volt power source. Both types of code readers

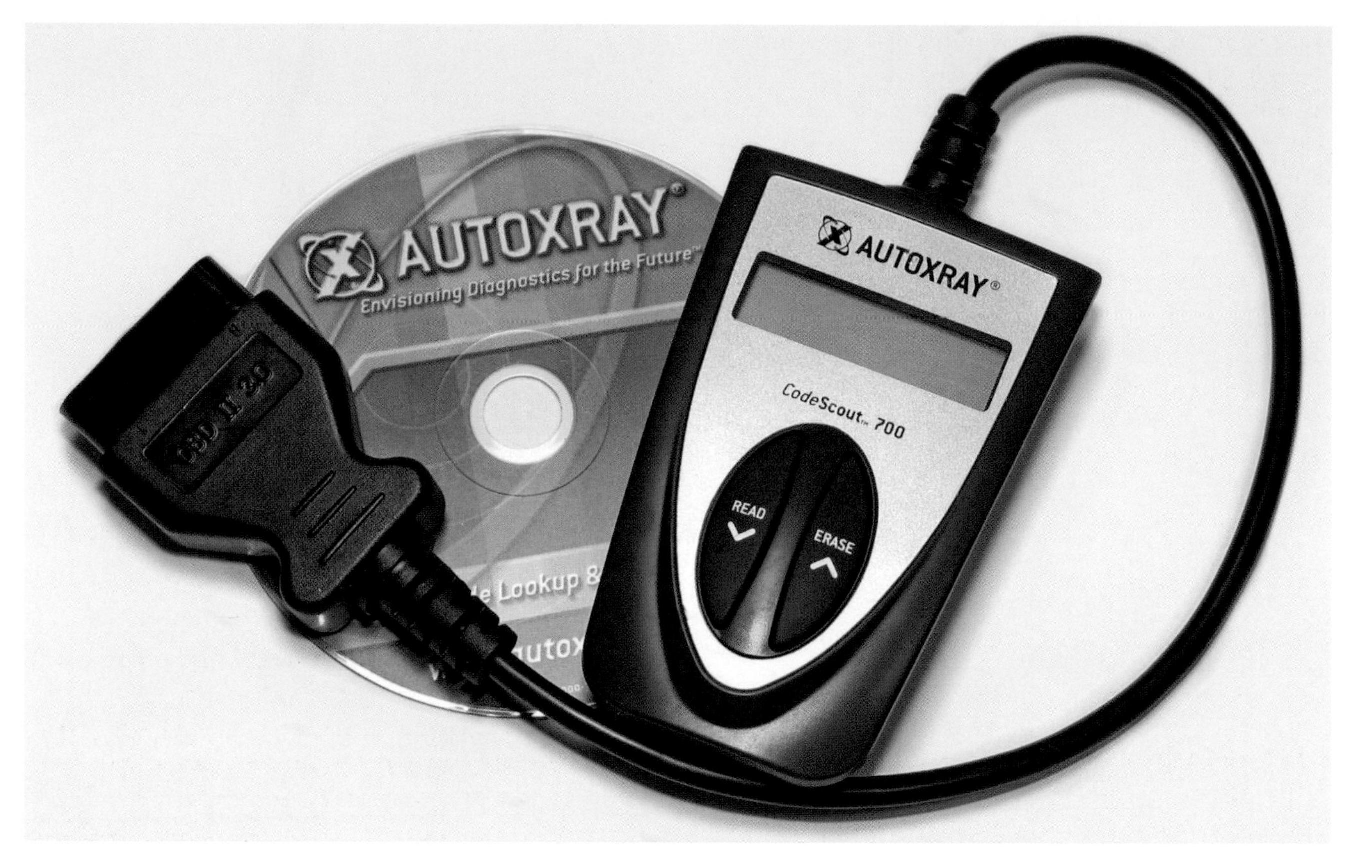

The CodeScout code reader from Autoxray is capable of reading all OBD-II communication protocols, including CAN. Simple two-button navigation makes this scan tool quite user-friendly and simple to operate for performing basic functions like reading and erasing diagnostic trouble codes or turning off a check engine MIL. *Courtesy Autoxray*

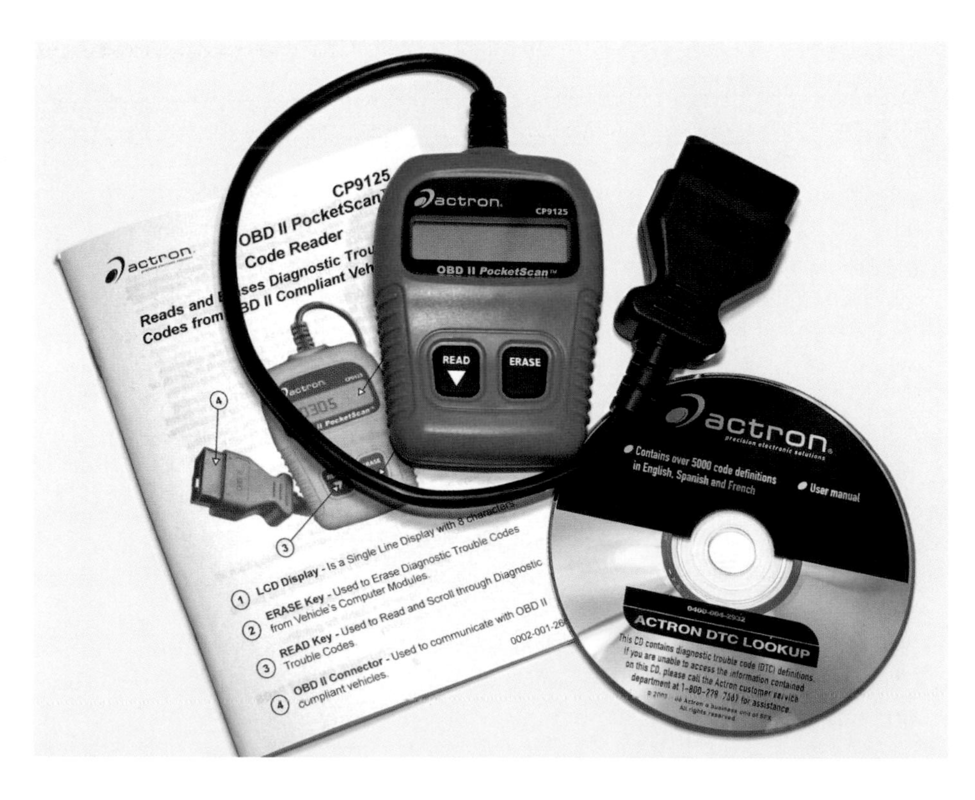

The OBD-II PocketScan comes with a CD containing an OBD-II trouble code library that provides definitions for generic DTCs that are capable of being read by this code reader. OBD-II monitor status can also be displayed, as can pending trouble codes and MIL status. *Courtesy Actron*

come complete with accompanying software, including a CD with DTC look-up software that includes manufacturer-specific "P1" codes, as well as an electronic user manual with generic OBD-II code definitions in English, Spanish, or French languages. Additional hard copies of user manuals in English and Spanish are also included. The CodeScout sells for around $80, while the PocketScan runs about $100, and both tools are available in major retail outlets and chain stores such as Sears, Wal-Mart, and most auto parts stores.

Following are specific steps for operating both the CodeScout and PocketScan code readers. By breaking these procedures down into sequential steps, a technician can gain a detailed look at what code readers can and can't do. The first step in using either of these code readers is to turn them on by plugging the OBD-II cable on the code reader into a vehicle's data link connector, or DLC. Once this is accomplished, a CodeScout's display will flash between "Code" and "Scout," while a PocketScan code reader will flash out "Pocket" and "Scan." If the "Read" key is pressed next, but the vehicle's ignition key has not yet been turned to the ON position, a "No Link" message will be displayed on the scan tool, which indicates the code reader is not yet communicating with the OBD-II system's PCM. However, if the ignition is in the ON position, but the scan tool still displays a "No Link" message, turn the ignition key to the OFF position for about 10 seconds, and then turn it back to ON again. Doing this should fix any interface communication problems between the vehicle's PCM and the code reader. Remember, the ignition key must be in the ON position to read PCM diagnostic trouble codes and also to check I/M readiness status.

With the scan tool ready to perform, a detailed description of how to read and erase diagnostic trouble codes when using either a CodeScout or PocketScan code reader is necessary. In addition, these scan tools have two functions—namely, checking MIL status and monitoring I/M monitors' operation—that will also be covered in more depth.

Reading Codes and Inspection and Maintenance Monitors

Once all of the uncompleted or "Incomplete" monitors have been displayed, the "Ready" monitors will be displayed next. Both code readers will display the actual number of monitors or tests the PCM has already successfully run and completed. In the example shown, the "3 Ready" display message indicates three successful monitors have been run by the PCM and are complete.

To read any OBD-II-generation vehicle's diagnostic trouble codes set by the PCM, simply press and hold the "Read" key on either code reader for about three seconds. An asterisk (*) symbol will move across the display panel while the scan tool is in the process of reading DTCs. If no trouble codes are stored in a PCM's memory, a "0 Codes" message will be displayed on the code reader's display panel.

2 Codes↓

The number of actual DTCs stored in a PCM memory will be displayed on the code reader's display screen, alongside an arrow that points downward, which is located to the right of the display. Pressing the "Read" key again will cause the scan tool to scroll the display downward to the first trouble code, and will then display the specific trouble code number on the display screen. Pressing and then quickly releasing the "Erase" key will cause the display panel to scroll through the entire list of trouble codes set by a vehicle's PCM.

P0304 

In the example shown, a specific P0304 (Cylinder 4 Misfire Detected) diagnostic trouble code is displayed on the code reader's display panel; as shown, the code reader reads the DTCs stored in a PCM memory and then displays them in this format on the code reader's display panel. Remember, there are no DTC definitions stored internally within these code readers, so the user manual will have to be consulted in order to determine what type of specific malfunction the code represents. Alternatively, this information is also available on DTC look-up software contained on a CD that accompanies each code reader.

P0514 Pd ↕

After the code reader displays the first DTC, by pressing the "Read" key again a technician can see the next specific trouble code set by a PCM as displayed on the code readers' display panel—in the example shown, it is a P0514 (Battery Temperature Sensor CKT Range/Perf) trouble code. The small letters "Pd" located on the right side of the display indicate that this trouble code is pending, which means the specified DTC malfunction occurred once during a single trip or drive cycle. However, in order for the same specific malfunction to actually cause a PCM to set a hard diagnostic trouble code and to simultaneously turn on an MIL, the same must occur a second time during a subsequent or second completed drive cycle. Consequently, the small letters "Pd" indicate the code reader is displaying a soft DTC (one that is merely pending) because the specific DTC may, or may not, eventually be set as a hard code by a vehicle's PCM (and subsequently stored in the PCM memory, causing the PCM to turn on an MIL light).

MIL ON ↕

After all DTCs have been read, pressing the "Read" key again will display MIL status—whether the MIL light is on or off. To obtain a correct reading for MIL status, a vehicle's engine must be running. While it may seem obvious that a malfunction indicator lamp on the instrument cluster is either on or off, a code reader can help determine if an MIL is really supposed to be on or not. In the graphic display shown, the code reader is displaying what a PCM is commanding an MIL to do. For example, if a code reader shows an MIL is on, but the light is not actually lit up on the dash panel, a malfunction somewhere in the MIL light circuit has occurred. The malfunction could be as simple as a burned-out bulb, but it could also be a more complicated MIL circuit wiring problem. The same conclusion would be drawn if the reverse were true as well—namely, if a code reader showed an MIL as being off, when in fact, the check-engine-soon light on the dash was actually lit up. Again, such a scenario would indicate some type of wiring problem in the MIL circuit, or possibly a malfunction in the PCM system is causing the MIL status on the instrument panel to disagree with the MIL status on the code reader.

Monitrs ↕

By pressing the same "Read" key again, the screen will display an abbreviated "Monitors" or "I/M Monitors" message. OBD-II monitors are actually just simple testing exercises run by a PCM to determine if all sensors and controls within a particular engine management system are operating properly. Either of the code readers we have been discussing will display the status of all monitors available on the specific vehicle that they are connected to.

2 Inc ↕

The first I/M monitors displayed on a code reader's display screen are all of those monitors or tests that have not yet been run by the OBD-II system's PCM. Later we'll take a look at monitors that have been run and are listed as ready. All monitor tests that have not yet been run by the PCM are listed as incomplete, which is typically abbreviated as "Inc." on the code reader's display screen, along with the total number of all such monitors not yet run. In the example shown, two monitors are listed as incomplete or not yet run by a PCM.

Misfire ↕

O2 Snsr ↕

Pressing the "Read" key again will cause the code reader to scroll down the monitor list, and eventually display which specific monitors or tests have not yet been run by the PCM. In this example, the "Misfire" and "O_2 Sensor" monitors have not yet been performed or completed. Such a scenario would most likely occur because *enabling criteria* (minimum threshold conditions that must be met before a monitor can be run or completed) for each specific monitor had not yet been met. Typically, just by operating the vehicle for additional drive cycles would eventually allow all monitor tests to run or be completed.

3 Ready

Comp

Catlyst

EGR

After the actual number of completed (and therefore "Ready") monitors has been displayed on a code reader's display screen, the next function the code reader will perform when the "Read" key is again pressed will be to display the names of the monitors (tests) that have been successfully performed and completed by the PCM and are therefore "Ready," or in a state of readiness. In the example shown, a comprehensive component monitor, catalyst efficiency monitor, and EGR monitor have all been successfully completed. The code reader is basically confirming that these systems are in a functioning, ready state, and that no malfunctions or DTCs exist in any of them. Continuing to press the "Read" key in succession will cause a code reader to scroll down through the list of successfully completed monitors. Once the last successful or complete monitor is reached, the arrow at the right of the display will now point up (meaning the bottom or end has been reached). When this occurs, the Erase key can now be used to scroll back up through all of the monitor displays as well as all the other displays. However, it is important to remember that the "Erase" key should be pressed and then quickly released in order to perform the scroll through the display function; otherwise, the code reader will try to erase any stored codes instead of scrolling.

Knowing which monitors have successfully been run by an OBD-II system's PCM and are now complete is quite helpful information, since some state's emissions' testing programs require certain, and sometimes all, monitor tests to be run and completed before testing can be passed. SAE guidelines suggest that all 1996 to 2000 model year vehicles should fail emissions' testing if three or more monitors have not been run and completed by a PCM prior to the state emissions' test being performed, while similar guidelines for 2001 and newer vehicle models mandate failure of state emissions' testing whenever two or more monitors do not indicate a "Ready" or complete status. See Chapter 2 for more information on OBD-II monitor readiness status.

Erasing Diagnostic Trouble Codes

ERASE?

In order to erase DTCs using these code readers and from a vehicle's PCM, make sure the vehicle's ignition key is in the ON position, then press and hold the "Erase" key on the code reader for more than three seconds. The display screen will display "ERASE?" To accept, simply press and hold the "Erase" key for more than three seconds again. The display panel will now show a dash moving across the screen. When the screen displays the message "DONE," all DTCs have been erased from the PCM memory. On many vehicles all the I/M monitors will be erased as well.

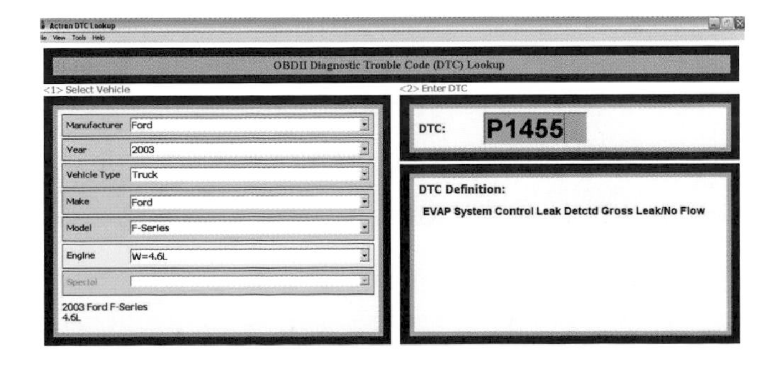

In the graphic shown, a manufacturer-specific P1455 DTC for a 2003 Ford truck is shown in the trouble code look-up software. This software is included with both the CodeScout and PocketScan code readers since these code readers do not have their own internal reference library from which to look up specific DTC definitions (whether the DTCs are manufacturer-specific or generic). *Courtesy Actron*

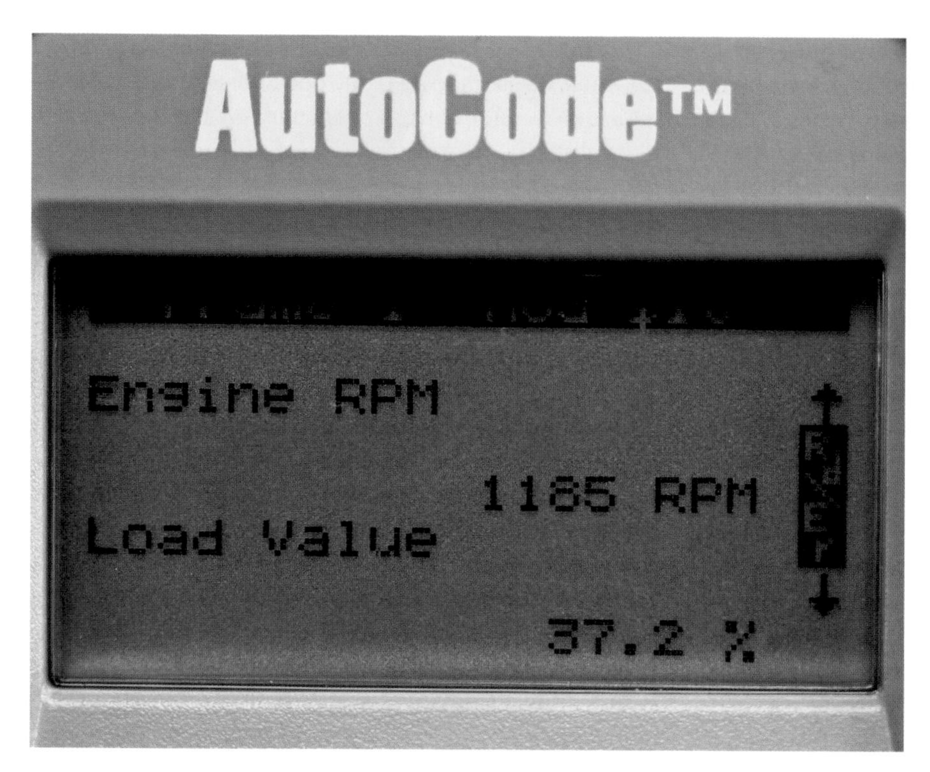

In the photo shown, when the specific P0113 (intake air temperature sensor malfunction) DTC was set by a PCM on this vehicle, the PCM also simultaneously recorded and stored freeze-frame data. Freeze-frame data can be displayed on an AutoCode reader screen for each corresponding DTC displayed. By using the down arrow key on the code reader, all of the freeze-frame data will scroll up on its display screen. More information on freeze-frame data can be found in Chapter 2. *Courtesy OTC Tools*

To erase codes from a vehicle's PCM using this code reader, select the "Erase Codes" function from the main menu. With the ignition in the ON position, simply press the "Erase" key on the code reader and then select "Yes" to erase all stored diagnostic trouble codes and freeze-frame data from the PCM memory. When the code reader has finished erasing all stored DTCs and freeze-frame information, a message that states "ERASE DONE" will display on the screen after the "Enter" key has been pressed.

Freeze-Frame Data and Other Functions

The majority of code readers can only read and erase a PCM's DTCs. However, the AutoCode reader is able to capture OBD-II freeze-frame data that the PCM simultaneously records whenever a DTC is set. When the "Freeze Frame" function is selected from the code reader's main menu, the specific DTC number is displayed on the first line of the display. The second line of text is freeze-frame data (see Chapter 2 for more information on freeze-frame data). Freeze-frame data is useful for determining the malfunction that caused a specific trouble code to be set, and stored in the PCM's memory. Freeze-frame data is a snapshot of all of the vehicle's sensor data at the time the DTC was set Depending upon a vehicle's manufacturer, freeze-frame data will typically be erased along with any DTC information when trouble codes are cleared from a PCM on most vehicles. Accordingly, it's usually a good idea to write down all DTCs and freeze-frame data displayed before erasing any DTCs or related freeze-frame information from a vehicle's PCM.

The AutoCode also permits a user to view MIL status, as well as status of I/M monitors whenever the function for any of these tasks is selected from the main menu. Checking MIL status enables a user to determine if the MIL is working properly. Remember, MIL status as displayed on the AutoCode reader should match MIL status on the vehicle in question (as determined by what the MIL is doing on the instrument panel or how it appears—on or off). In addition to MIL status, I/M monitors can be viewed on this code reader by selecting that function from the main menu. Depending upon vehicle manufacturer, the status of as many as 10 monitors is available for viewing. The status of any I/M monitors will be indicated as "Ready" once a monitor has been run and completed by the OBD-II system, or "INC" (for incomplete status) if the vehicle was not driven enough to complete a specific requisite drive cycle necessary to run a particular monitor, or "N/A" if the vehicle does not support the performance of a specific monitor or test. For example, if a secondary air monitor is listed as "N/A" it means the specific vehicle being scanned does not use that particular emission system.

A "State OBD Check" function is just one more of the many enhanced features found on the AutoCode reader that can be selected from this tool's main menu. The data displayed during a state OBD check provides basic status information concerning a vehicle's OBD-II system, including MIL status (on or off), number of DTCs found, and number of test monitors available as well as their current status (ready, incomplete, or not applicable). This test function was primarily designed for quick and easy verification of the status of all OBD-II system functions and components. (It is typically used during state emissions testing—hence its name.) Bear in mind, this test should be performed while the engine is running so MIL status can be correctly displayed. Remember, however, that the number of codes displayed on the panel are only OBD-II generic codes and are not pending codes that could show up after completion of several additional drive cycles.

Another great feature found on the AutoCode scan tool is its internal DTC look-up library. When selected from the code reader's main menu, the Code Lookup function allows users to enter both generic and manufacturer-specific DTC numbers into the code reader in order to obtain SAE standardized definitions of any DTCs entered. Because the AutoCode also has internal batteries, users can view the trouble code library even when the tool is not connected to a vehicle. This feature is particularly useful when DTCs have been erased during a previous code scan, but a user wants to see definitions for codes without having to pull up stored codes or having a particular trouble code present. This feature also saves users time that would have otherwise been spent looking up codes in the manual or on the Internet.

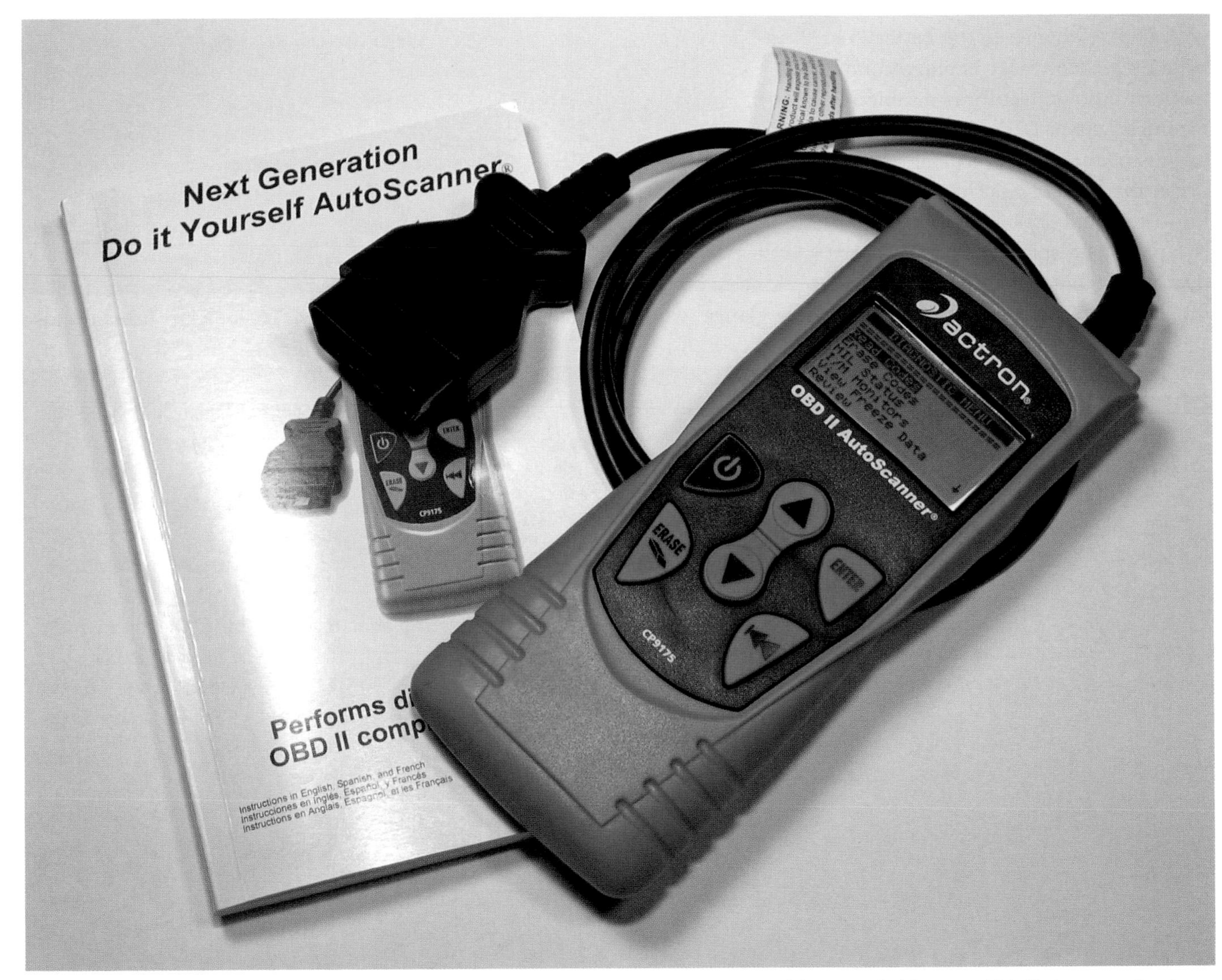

In addition to reading and clearing DTCs, an Actron OBD-II AutoScanner code reader can store data generated from a vehicle that was previously scanned, and this information can be viewed even when the code reader is not connected to a vehicle. Furthermore, this tool can be programmed by users to display information in English, Spanish, and French. *Courtesy Actron*

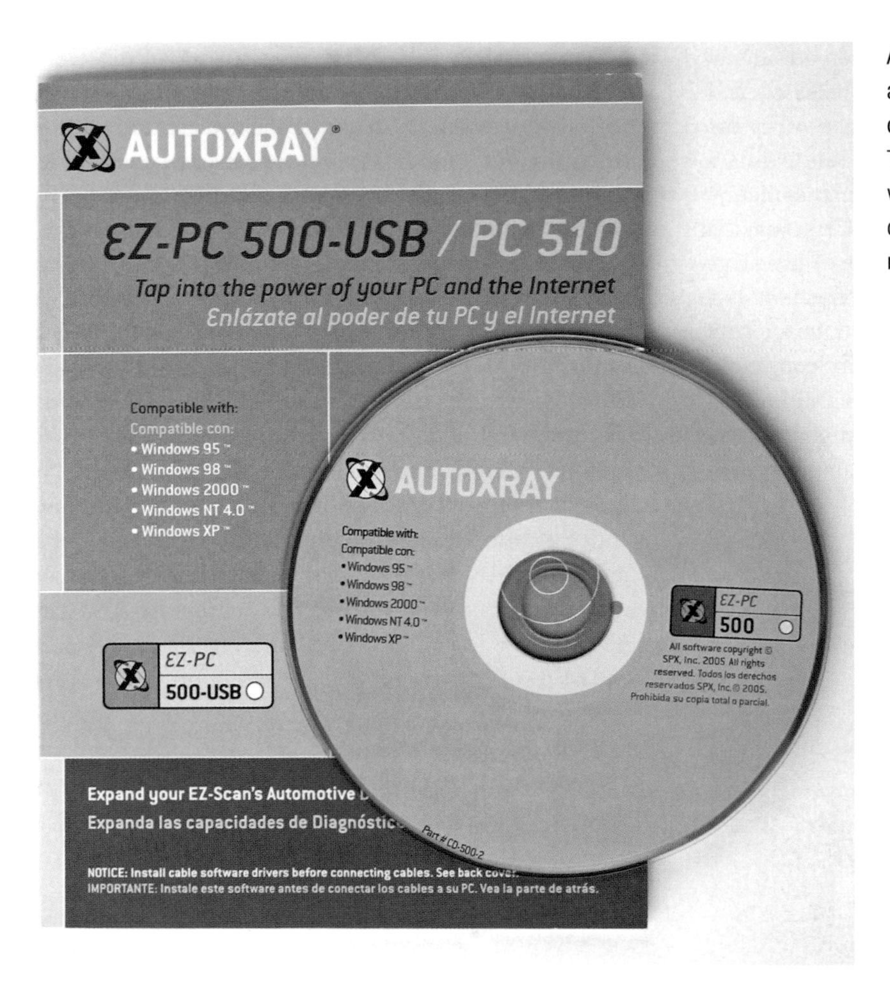

AutoXray also produces computer software known as EZ-PC 500-USB, which can be used in conjunction with any AutoXray EZ-Scan scanner. This software package allows any captured vehicle data to be uploaded to a PC or laptop. Data can also be stored and saved as a text file for later reference. *Courtesy AutoXray*

SCAN TOOLS

(milliseconds, or one second). However, capture time lengths of between 500 milliseconds (1/2 second) and as much as 5,000 ms (five seconds) can be set. By specifically setting capture rates or lengths of time for the data frame (during which data can be recorded), a user can utilize this scanner to zero in on specific drivability problems or to capture broader ranges of data for more generalized diagnoses of drivability problems.

The useful capture data graph found in the EZ-PC computer software program from AutoXray (Figure 5-13 on p. 107) can be used to overlay data parameters and then view them on the same graph. For example, the data from an upstream oxygen sensor (O2S11) could be compared to the data from a downstream oxygen sensor (O2S12) to determine how efficiently the catalytic converter is operating during a test drive. In addition, throttle position sensor data can be graphed along with vehicle rpm and calculated load data in order to locate a faulty throttle position sensor. Using EZ-PC software is a simplified method for verifying when repairs are successful, since data from before and after a specific repair can be compared.

Elite AutoScanner Pro

The Elite AutoScanner Pro scan tool is at the top end of the do-it-yourself aftermarket products since it offers more features and functions than many professional scanners did only a few years ago. This near-professional-grade tool features a large 2 5/8 x 1 5/8-inch high-resolution display that's easy to read, and when connected to a vehicle it's backlit for easy nighttime viewing or use inside a dark garage. In addition, this tool can be specifically programmed by a user according to preference for English, Spanish, or French. Other user-programmable features include a specification for the length of time before the tool automatically shuts off, a keypad on/off beeper alert, and custom user key programming. Priced at around $550, these features make this tool a competitive choice for diagnosing electronic vehicle malfunctions,

A USB cable allows the tool to be connected to a PC or laptop where recorded vehicle data stored as a text file for later reference or printing. All PIDs can be recorded or the user can select specific PIDs that pertains to diagnostic codes present in the vehicle that the tool is connected to. The Elite

EZ-PC - [Capture Data (List)]

File Software Update Tool Link Window Help

Parameter Name	-C	-B	-A	Trigger	+A	+B	+C
Fuel System 1 (Closed Loop=0)	0	0	0	0	0	0	0
Fuel System 2 (Not Supported=0)	0	0	0	0	0	0	0
Load Value (%)	60.30	60.30	19.90	19.90	27.30	27.30	27.30
Coolant Temp (°F)	179.00	179.00	179.00	179.00	179.00	179.00	179.00
Short Term Fuel Trim1 (%)	2.30	2.30	7.00	7.00	0.70	0.70	0.70
Long Term Fuel Trim1 (%)	2.30	2.30	2.30	2.30	2.30	2.30	2.30
MAP Sensor (in.hg)	17.00	17.00	27.00	27.00	27.00	27.00	27.00
Engine RPM (RPM)	1734.00	1734.00	1734.00	3276.00	3276.00	3598.00	3598.00
Vehicle Speed (MPH)	8.00	8.00	8.00	19.00	19.00	29.00	29.00
Ignition Timing Adv (°)	36.50	36.50	36.50	28.00	28.00	28.50	28.50
Intake Air Temp (°F)	87.00	87.00	87.00	87.00	87.00	87.00	87.00
Throttle Posn (%)	11.00	11.00	11.00	56.00	56.00	54.00	54.00
Oxygen Sensor 1 Bank1 (V)	0.52	0.52	0.52	0.52	0.52	0.54	0.54
Short Term Fuel Trim1 (%)	3.00	1.00	1.00	1.00	1.00	2.00	2.00
Oxygen Sensor 2 Bank1 (V)	0.81	0.28	0.28	0.28	0.28	0.28	0.78
Short Term Trim2 (Not Supported=0)	0	0	0	0	0	0	0
OBD II Require (OBD II (CARB)=0)	0	0	0	0	0	0	0
Battery Voltage (V)	13.90	13.80	13.80	13.80	13.80	13.80	13.70
Throttle Sensor (Volts)	1.00	1.74	1.74	1.74	1.74	1.74	2.44
Front O2 Sensor (V)	0.83	0.75	0.75	0.75	0.75	0.75	0.22

Ready

Figure 5-12. Here is a sample Capture Data List graph produced by AutoXray's EZ-PC computer software program. Note that each column header represents specific frames of captured data, with the trigger point in the middle of the chart. Thus, the data stream from immediately before and after the trigger point can be viewed and analyzed. Each of the various PID are listed in the far left column. The vehicle speed has been highlighted in gray for easier viewing. *Courtesy AutoXray*

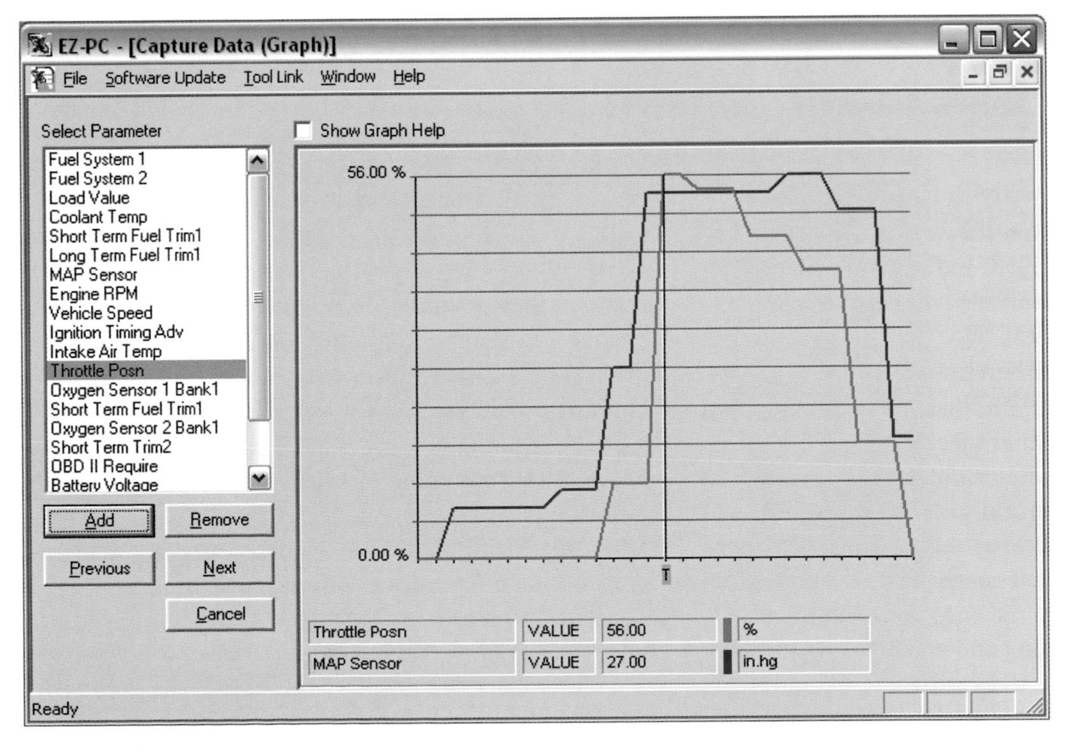

Figure 5-13. Captured data from the EZ-PC computer software program shows throttle position (green graph trace) as it relates to MAP sensor values (red graph trace)—as many as eight PID can be displayed on a graph page at any one time. The small lines at the bottom of the graph represent data frames, while the "T" in the middle represents the trigger point for this particular sample recording. *Courtesy AutoXray*

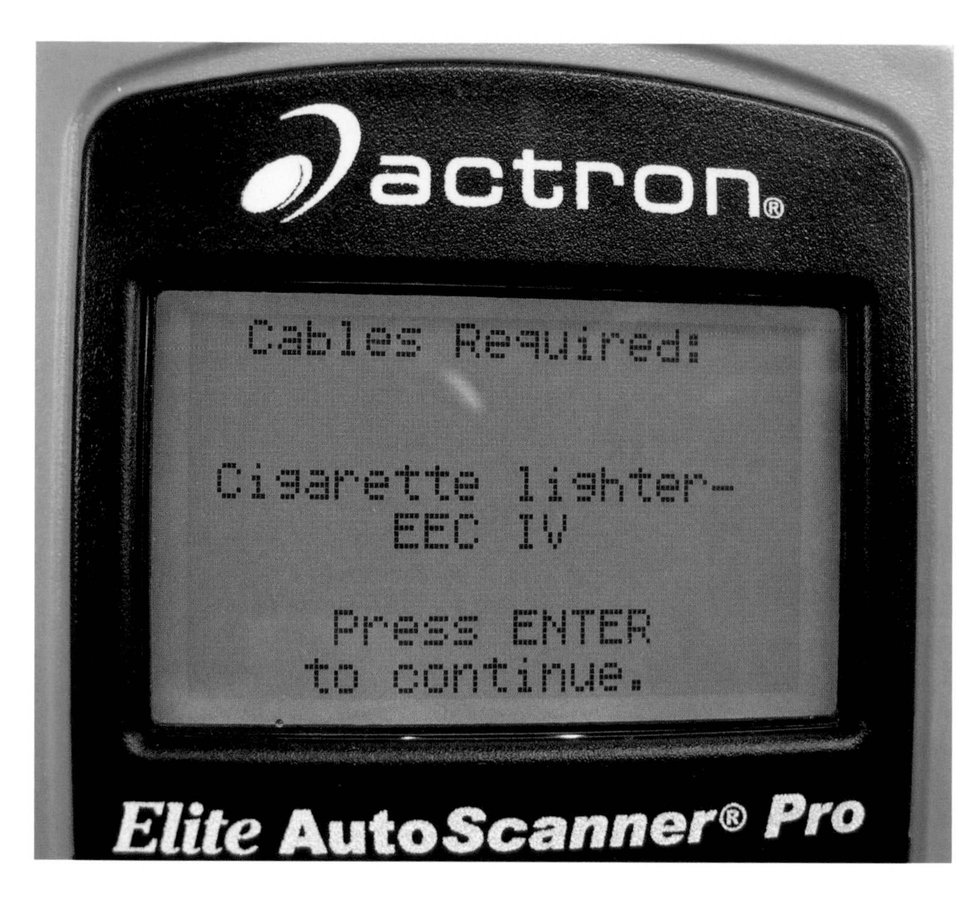

Some OBD-I-generation vehicles have specific cable connections. Fortunately, the Elite AutoScanner Pro scan tool pictured is capable of providing a user with instructions required to connect an OBD-I-equipped vehicle directly to the scan tool. The full instructions are actually displayed as text on the screen. The sample display screen pictured shows that an external 12-volt power supply is required in order to properly connect a Ford Motor Company EEC-IV vehicle to the scanner. *Courtesy Actron*

enhanced manufacturer-specific DTCs for General Motors, Ford Motor Company, Chrysler, Toyota, and Honda vehicles. In addition, certain manufacturer-specific trouble codes can also be displayed for several Asian and European vehicles. This tool is priced at around $570, but is well worth the price, since it can also monitor the performance of drivetrain-related components and is compatible with all OBD-II system protocols, including CAN.

This scanner also has a large screen that enables it to display both DTCs and their respective EPA-approved code definitions at the same time. Once the scanner is plugged into a vehicle, the 128x128-pixel LCD screen is immediately brightly backlit, making viewing in a dark shop or under the hood at night possible. In addition, the display screen has a user-adjustable contrast function that makes viewing the screen in bright sunlight equally possible.

All compatible software is menu-driven and user-friendly. Plus, this scan tool can perform many functions, including I/M readiness status and testing, reading and erasing DTCs, freeze-frame data capture and display, diagnostic monitor tests, O2 monitor tests, OBD-II drive cycle modes, identification of modules present, and review of all data. The ScanPro scanner also has a Flight Record and Playback feature that allows simultaneous recording of live data streams from a vehicle's PCM. The PIDs can also be displayed in a graph format, and since the display screen is so large, additional data can be displayed underneath all graphs, enabling users to view additional information in conjunction with the graph. Whenever recorded data is played back in graph mode, a cursor appears below the graph; this cursor can be moved along the time axis line of the graph, while the corresponding numbers that are represented in the graph are displayed below. The use of graph format data enables users to much more easily and effectively diagnose front and rear O2 sensor operation, as well as the operation of other sensors.

The ScanPro 3409 also comes complete with its own soft carrying case and detachable OBD-II cable for connecting the tool to vehicles. The heavy-duty OBD-II cable is 7.5 feet long, making it easy to operate the scanner while working under the hood of a car or truck. In addition, a serial cable (purchased separately) can be used to connect the scanner to a PC or laptop, where software updates are available via the Internet. The serial cable can also connect the scanner to a printer, allowing the tool to be programmed for a variety of printer configurations. A 120-page user manual is also contained on the CD in a PDF that can also be printed out. Internal batteries allow data to be viewed from off-vehicle locations.

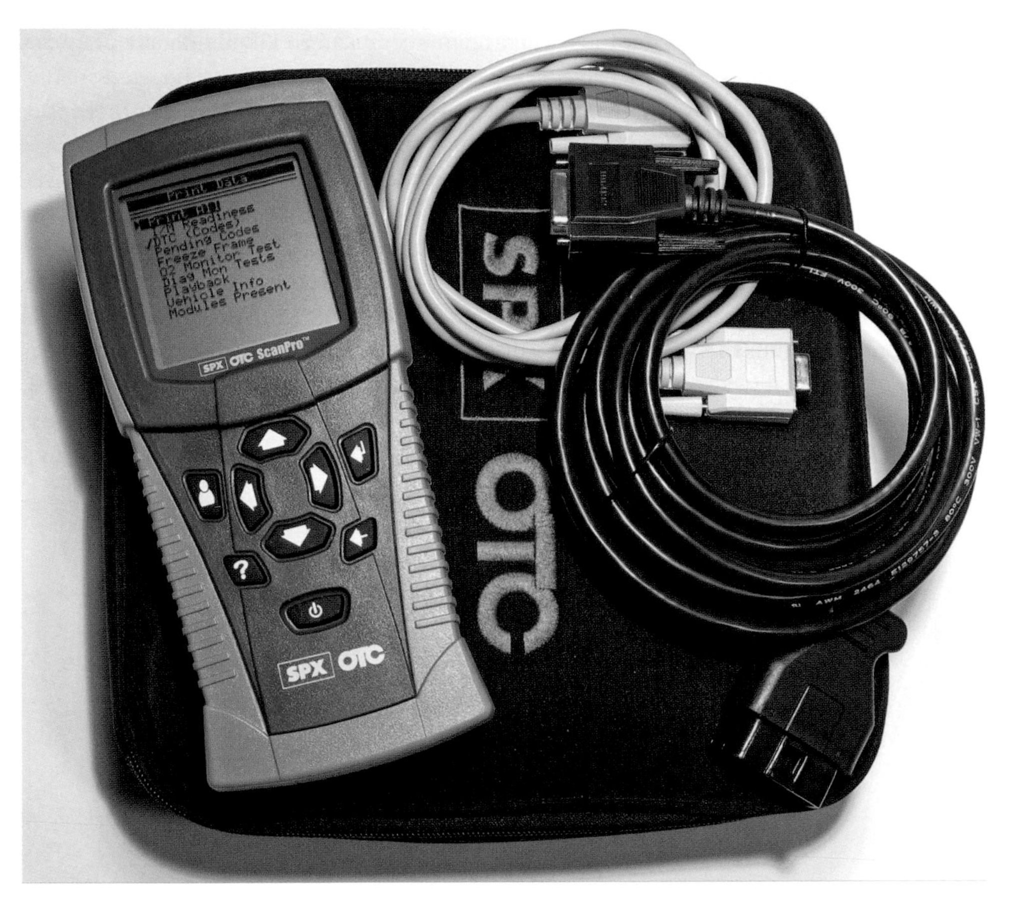

Pictured is an OTC ScanPro 3409 scan tool. The large-screen format is capable of displaying large amounts of information from OBD-II system vehicles, including live data streams and graphs. In addition, a flight record and playback feature allows data to be captured during vehicle testing. All data can be uploaded to a PC or laptop for storage or subsequent viewing. *Courtesy SPX/OTC Corporation*

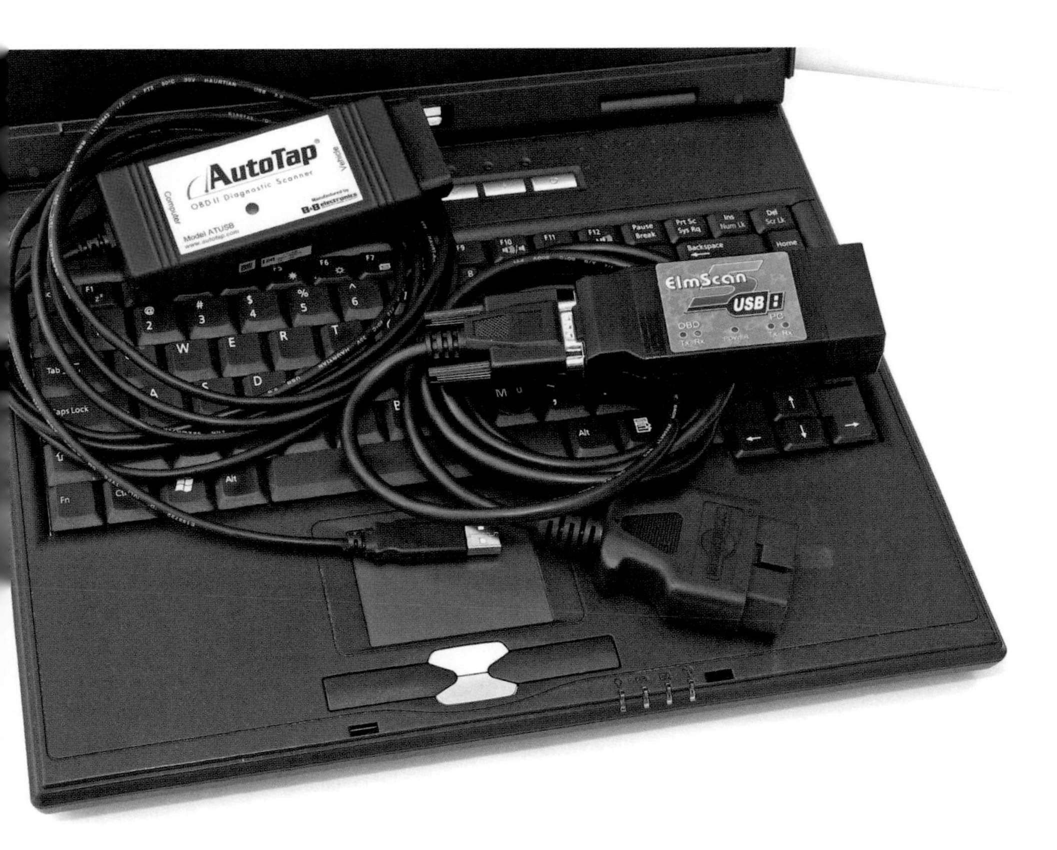

Most people don't realize it, but if they own a laptop they already have one-third of a scan tool. Simply by adding appropriate diagnostic computer software and a hardware OBD-II interface, a laptop can become a powerful portable scanner with a large, high-resolution display screen. As an example, AutoTap and ElmScan hardware and cabling are shown in the photo. With the respective software loaded onto a laptop, a user-friendly scan tool is created. *Courtesy AutoTap and ScanTool.net*

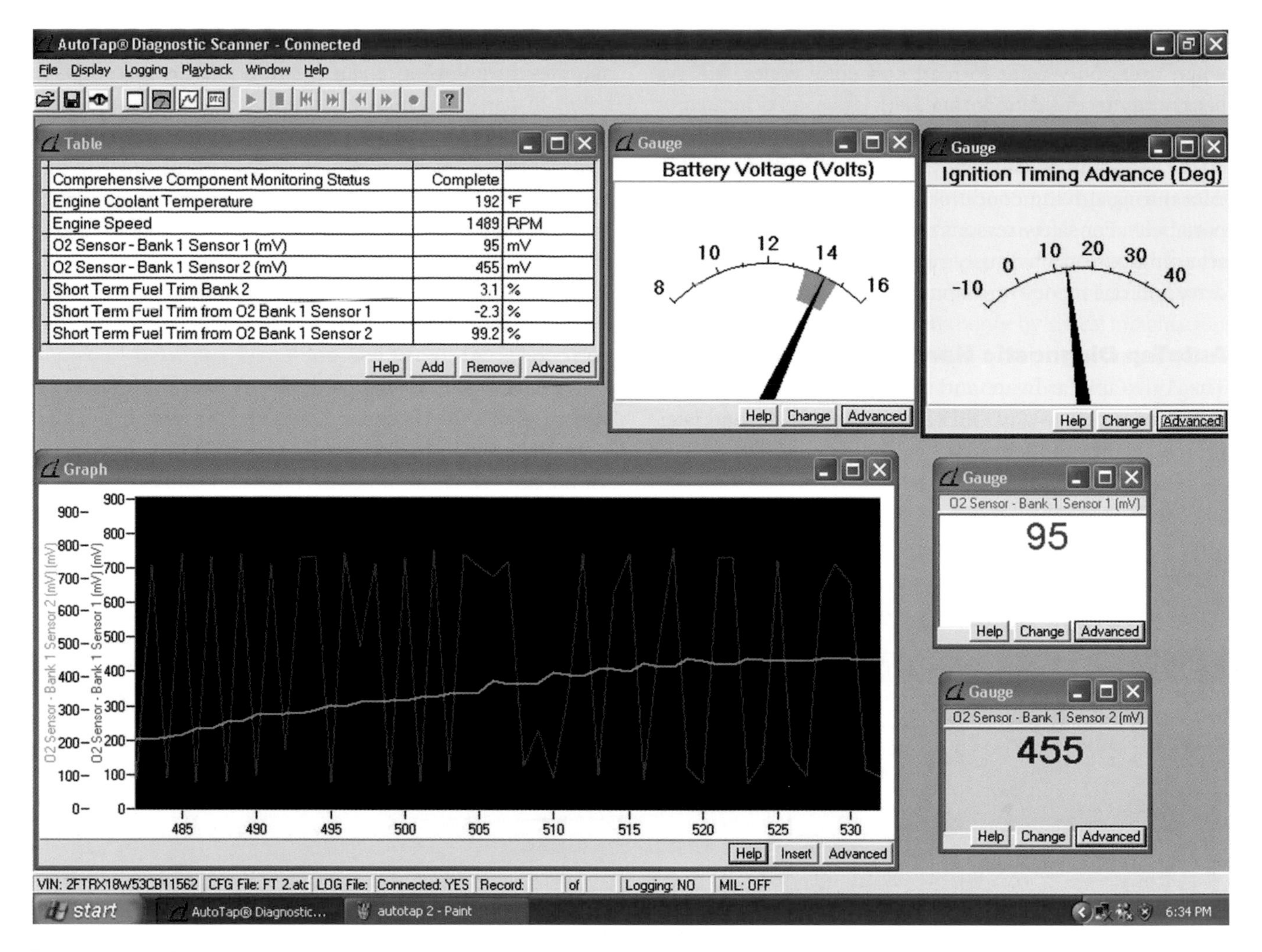

The sample AutoTap scan tool data configuration display screen shown represents a configuration programmed to diagnose front and rear O2 sensor performance. The data parameters are displayed in the chart or table (top left), along with battery voltage and ignition timing advance gauges (top, middle, and right, respectively) in degrees. The electronic graph displayed (on bottom left) shows performance data over time for the Bank 1 (left side of engine), upstream oxygen sensor (O2S11) in red, and the same information for the Bank 1 (left side of engine) downstream oxygen sensor 2 (O2S12) in green. Digital values for each respective sensor are also shown (bottom on right). In this specific instance, it is readily apparent that both oxygen sensors are working normally, as shown by the graph, since the upstream O2S11 sensor's red line data stream shows this sensor is switching rapidly (visually up and down) as the PCM controls fuel distribution in normal closed-loop mode, while the O2S12 downstream oxygen sensor's green line data stream is holding steady, indicating the catalytic converter is storing and releasing oxygen properly and otherwise operating correctly. *Courtesy AutoTap*

The ElmScan scanner from ScanTool.net connects between any OBD-II-compliant vehicle and a laptop or PC. ElmScan comes with ScanTool.net software, but it can also run various "freeware" available on the Internet. *Courtesy ScanTool.net*

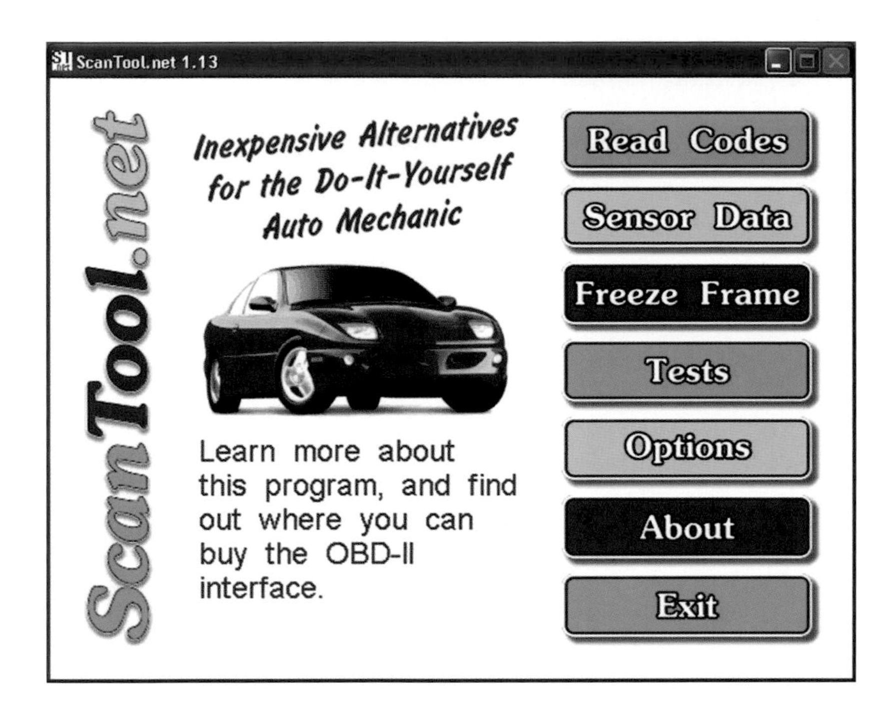

This picture shows a sample Main Menu electronic display page from OBD-II system software manufactured by ScanTool.net. As can be seen, the user interface is simple and easy to use. Currently available features include Read Codes and Sensor Data. Future releases plan to include data graphing and data logging, reading freeze-frame data, and tests for continuous and noncontinuous oxygen sensor test results. *Courtesy ScanTool.net*

users the ability to access enhanced parameters, including those from General Motors, Ford Motor Company, and Chrysler.

Each of the various displays provides general information to users, including the reading and clearing of DTCs (specific EPA code definitions are also displayed on-screen), the ability to view/diagnose a bad O2 sensor, live data from the manifold absolute pressure (MAP) sensor and the mass airflow (MAF) sensor, engine coolant temperature (ECT), and live data from a throttle position (TPS) sensor as well as other numerous sensors under a PCM's control. In addition, useful customizable virtual dashboards can be created using both analog and digital gauges, graphs that show single or multiple data streams, and charts or tables that list data parameters and/or the status of the OBD-II system. Furthermore, 11 OBD-II monitors can also be viewed to determine EPA readiness status, as can freeze-frame data. All data can be recorded and saved for later off-vehicle viewing.

ElmScan

The ElmScan 5 Scan Tool is an interface hardware package that has the capability to connect any PC or laptop to any 1996 or newer OBD-II-compliant vehicle. This scanner is based on an ELM 327 processor, and is available in three different configurations: serial port, USB cable, and Bluetooth wireless. The serial port version costs approximately $115, while the USB cable version typically runs around $130, and the wireless version is usually priced around $200. The ElmScan 5 supports all OBD-II protocols, including CAN, and is compatible with a number of software applications, including open source software or shareware. Numerous free software programs work in conjunction with the ElmScan 5, as will many other software programs available for purchase on the Internet. The ElmScan 5 comes with specifically designed compatible ScanTool.net OBD-II software, which is available in two operating system platforms—either DOS or Windows. Minimum PC/laptop system requirements are as follows:

DOS Operating System

- 386SX 10Mhz (megahertz) processor or higher
- 1Mb RAM
- DOS version 3.0
- 640x480-pixel display screen
- Serial port

Windows Operating System

- Windows 95, or Windows 98, NT, 2000, Me, or XP
- 486DX 25Mhz processor or higher
- 4Mb RAM
- 640x480 display with DirectX 7
- Serial or USB port

PDA Interface Scan Tools

For those who own a personal digital assistant (PDA) electronic device or a pocket personal computer, there are hardware components and software programs available to connect these potential scan tools to an OBD-II compliant

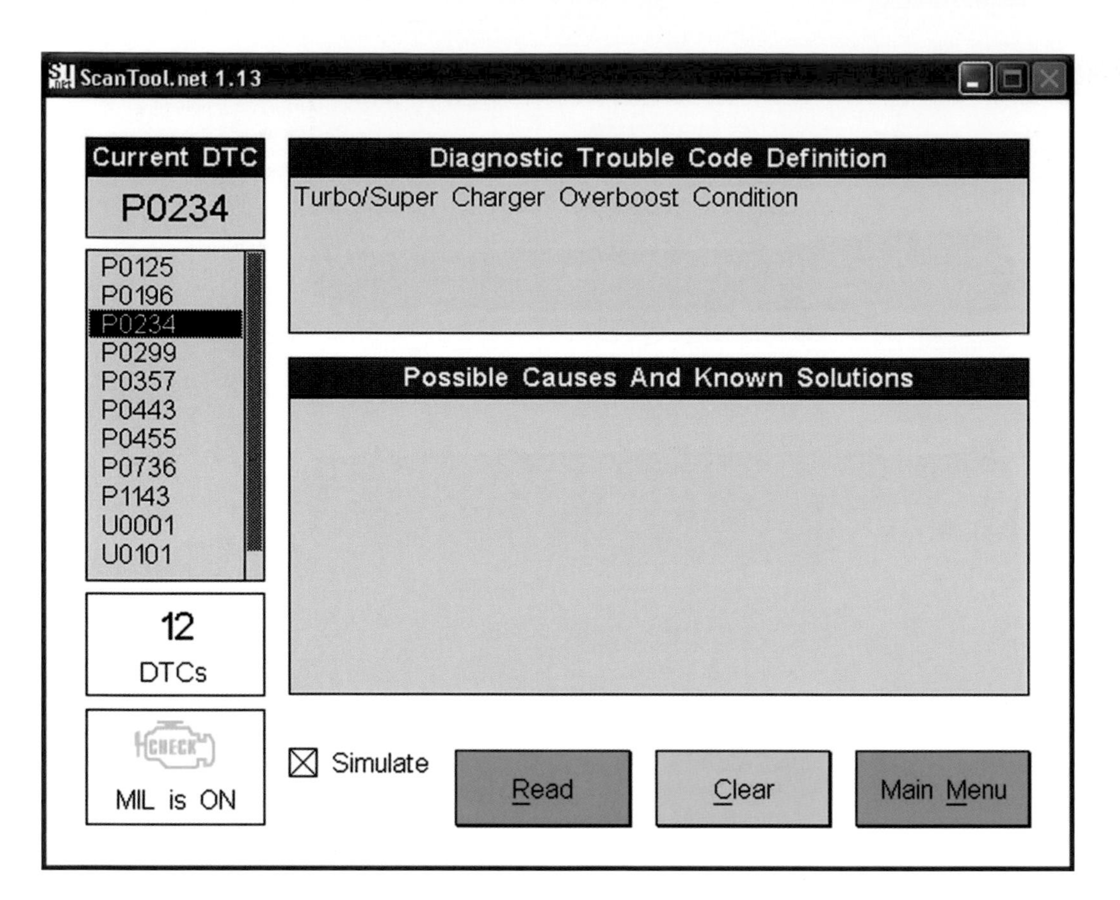

Pictured is a sample display screen from an ElmScan 5 scan tool. As can be seen, this scan tool displays OBD-II system DTCs, along with associated specific EPA standardized trouble code definitions and MIL status. *Courtesy ScanTool.net*

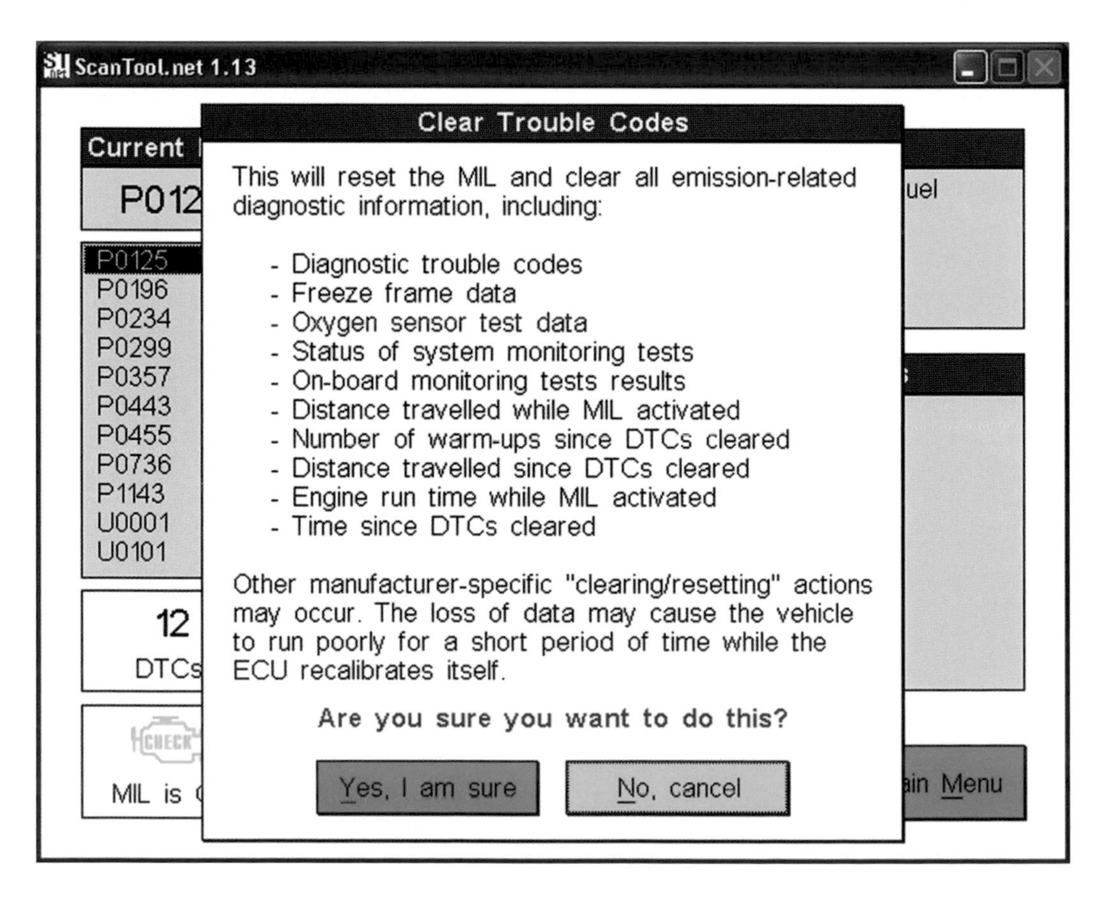

This picture shows another sample display screen from an ElmScan 5 scan tool that tells a user how DTCs can be cleared and an MIL turned off. Specifically, the "Clear Trouble Codes" warning screen is shown. *Courtesy ScanTool.net*

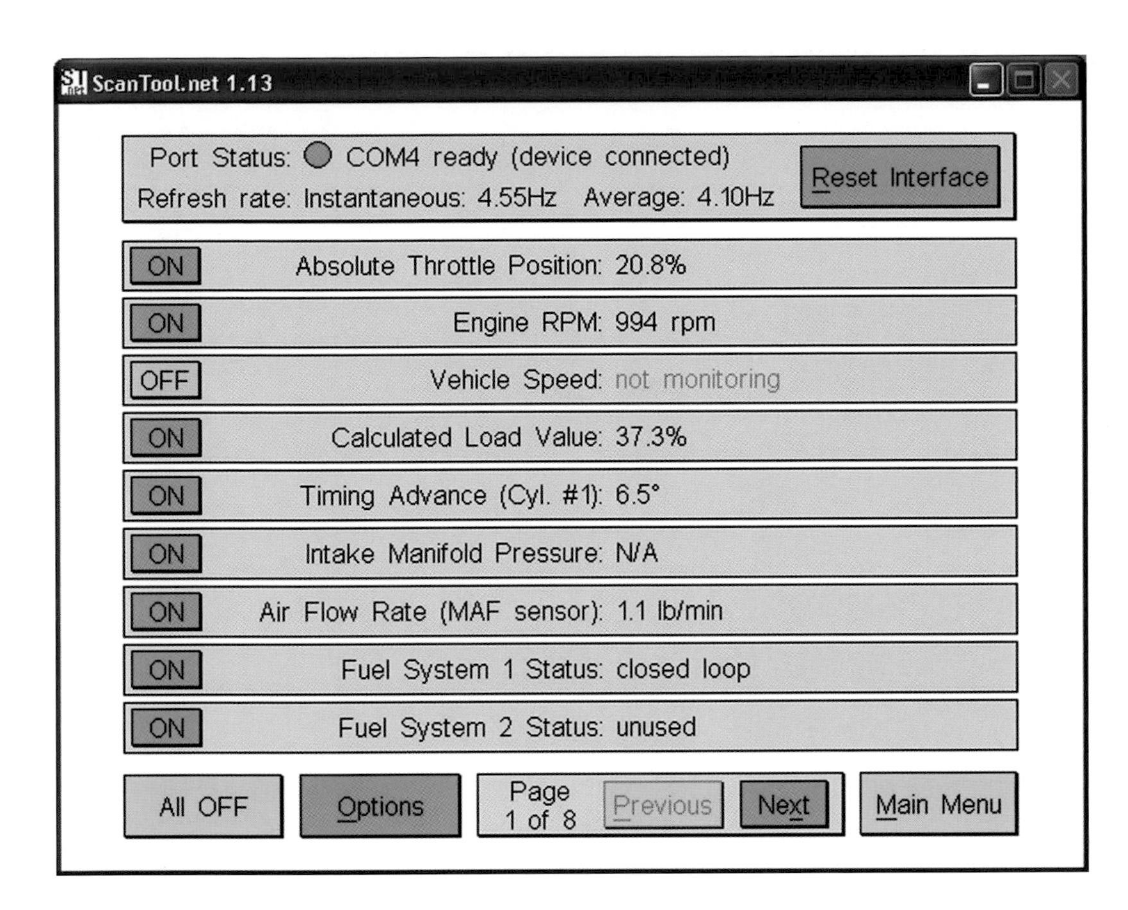

Shown in the picture is a sample of a ScanTool.net sensor data screen that displays generic OBD-II parameters or PID. These data parameters can be turned either on or off—the fewer that are on, the faster the refresh rate. *Courtesy ScanTool.net*

EASE Diagnostics sells hardware and software necessary to connect a pocket PC or PDA device to an OBD-II vehicle. Pictured is their WIN CE/Pocket PC Scan Tool Package. The HP iPAQ pocket PC (upper right corner) is not included in the package, but is also available as well from this same manufacturer's website. *Courtesy EASE Diagnostics*

vehicle. EASE Diagnostics sells hardware interface cables for all types of PDA devices. With additional software programs available for these devices, test results and vehicle scan data can be uploaded to a PC for storage or subsequent viewing. In addition to reading and clearing DTCs, these devices can also display data in chart or table, meter, or graph formats. Screen menus are easy to navigate, as they feature familiar Windows icons and drop-down menus. The combined hardware and software package for generic OBD-II vehicles typically costs around $500, with additional manufacturer-specific enhanced software available for an extra cost.

The combined EASE WIN/CE Pocket PC scan tool hardware and software package works well with most pocket PC devices that are capable of running Windows CE 3.0 or higher, or Pocket PC 2002 or Pocket PC 2003 operating systems. Various computer manufacturers supported include Hewlett-Packard, Compaq, Audiovox Maestro, Intermec, Symbol, Toshiba, Casio, and Dell Axim. Manufacturers of PDA devices supported include Palm, IBM Workpad, HandEra, Handspring, and Sony. For further information, the EASE diagnostic website lists specific platforms at www.obd2.com. EASE diagnostics also manufactures compatible EASE PC scan tool software for use in PCs and laptops.

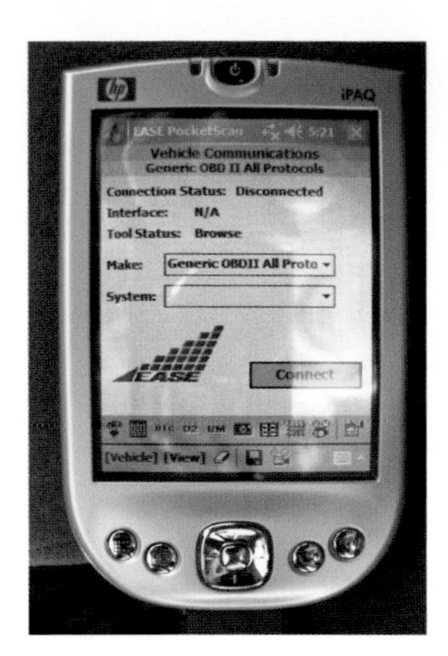

After connecting the cable from an OBD-II system to a vehicle and then powering up this HP iPAQ pocket PC, the EASE diagnostic program will launch. This software program is menu-driven and uses Windows drop-down lists and icons to navigate between various screens. *Courtesy EASE Diagnostics*

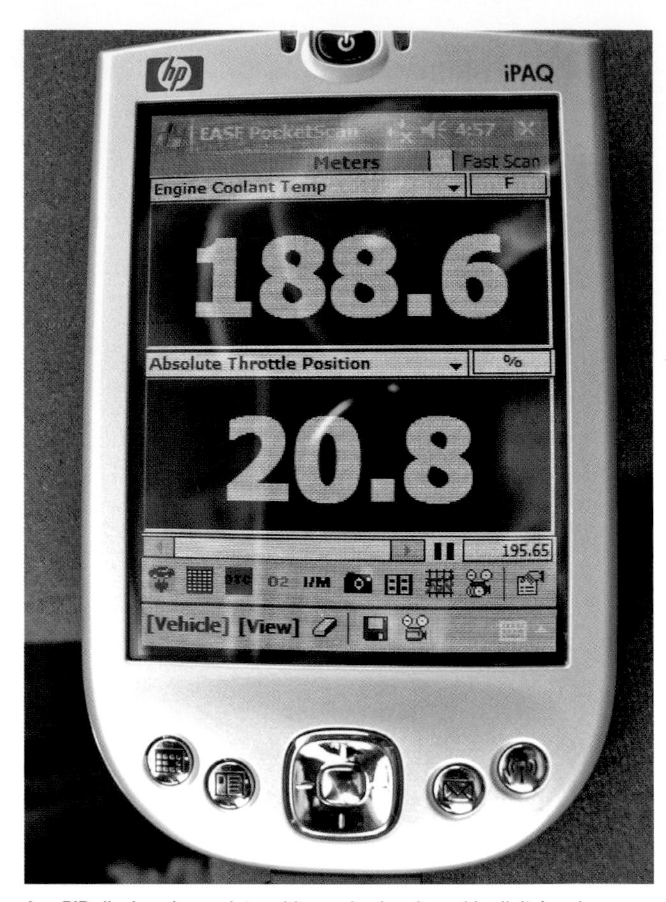

Any PID displayed on a data grid can also be viewed in digital meter format. In this format, important information can be viewed at a glance, enabling simultaneous operation of a vehicle. All data can be displayed in both metric and standard formats. *Courtesy EASE Diagnostics*

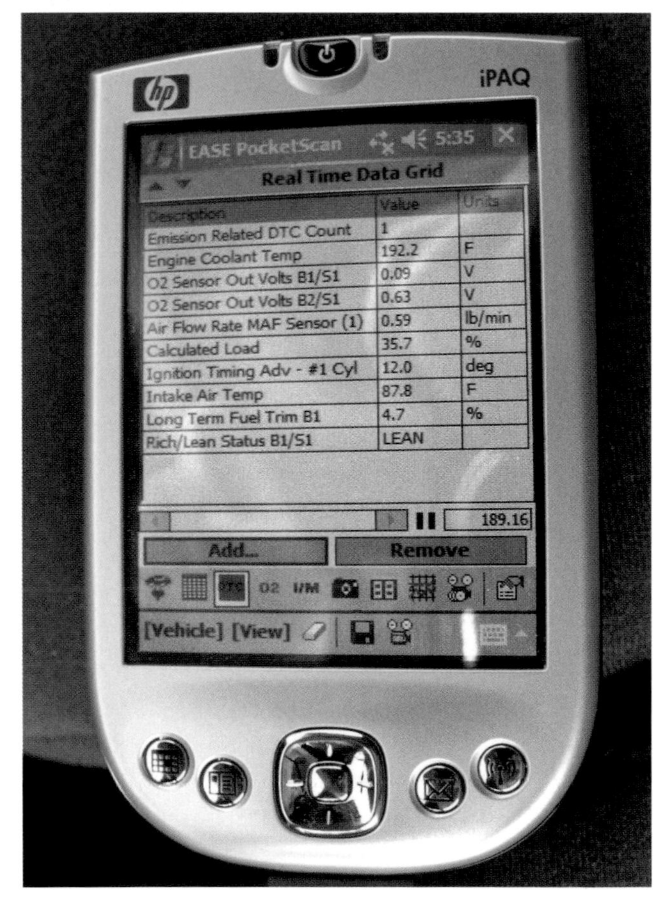

Shown in the photo is a Real Time Data Grid from an HP iPAQ pocket PC; this informational grid can be customized by adding or removing data parameters. In addition to displaying data, this computerized scanning system also maintains a trouble code library that allows users to look up EPA standardized DTC definitions. In addition, O2 sensor data and tests are available for all later model OBD-II vehicles, as are I/M test results. Moreover, whenever a DTC is set, simultaneous freeze-frame data can be viewed and saved as well as all other data. *Courtesy EASE Diagnostics*

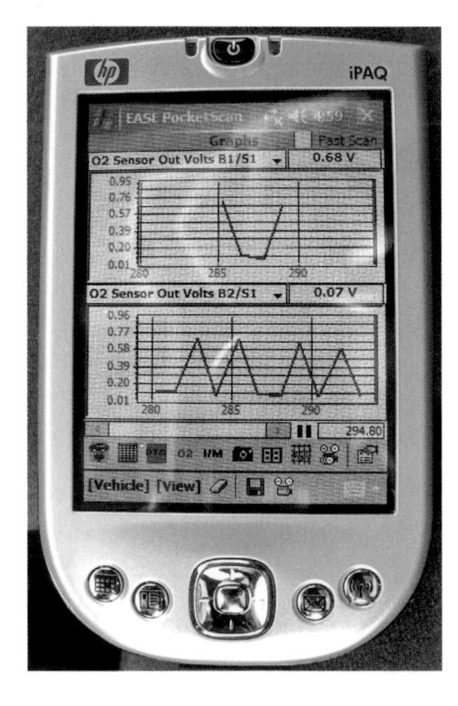

Any data added to a real-time data grid can be displayed in graph format. Graphs can display live or recorded data. For example, digital readouts are shown in the photo, along with a graph format display of real-time data (X and Y axis scaling is automatic). *Courtesy EASE Diagnostics*

CHAPTER 6
AUTOMOTIVE DETECTIVE WORK

The same mechanical problems this vintage in-line six-cylinder engine on this old Ford was subject to (when it used to run) still occur in contemporary late-model cars and trucks. Both early- and late-model engines use pistons, rings, valves, a head gasket, and intake and exhaust manifolds that all wear out or leak as miles rack up. The tried-and-true techniques that have worked for years to diagnose common engine mechanical problems still work today. Furthermore, when these techniques are applied to an engine on a late-model vehicle, they can save lots of time that would otherwise be wasted chasing false OBD-II DTCs that are triggered by engine mechanical problems. *Courtesy Elwoods Auto Exchange*

While this book focuses on onboard computer systems (OBD-I and -II), code readers, and scan tools, it's important to remember that what powers the majority of automobiles and light trucks driven today is four-stroke, gasoline-powered engines (apologies to owners with diesel-powered vehicles). While OBD-II systems provide a wealth of information about emissions-related components that are operating to reduce exhaust emissions, the same cannot be said for mechanical engine problems. onboard computers cannot always set trouble codes for engine mechanical problems. Although OBD-II diagnostic monitors all relate in some way (and perform in response) to what's coming out of an engine's exhaust, it is a simple limitation that, despite all their sophistication, these systems still rely on engines that can optimally burn air/fuel mixtures properly. Surprisingly, even if something serious is out of whack with basic mechanical engine operation, a PCM still may not be able to figure out the problem in order to set appropriate DTCs.

Most professional technicians know that many drivability problems and numerous DTCs often do not actually reflect high-tech malfunctions (as a DTC might indicate); something more basic that is of a mechanical nature that the onboard computer system simply may not be able to detect or set a proper (or any) trouble code for could actually be the problem. For example, when an engine misfires, a No. P0300 multiple misfire DTC—or even a code indicating a specific cylinder misfire—could be triggered. The misfire might simply be caused by a mechanical engine problem rather than an electronic malfunction in an OBD-II system component (ignition or fuel injection system). In such a case, a PCM is simply not able to differentiate the cause, source, or nature of the problem. A technician could spend hours with a scan tool or code reader trying to figure out an electronically related reason why numerous DTCs have been set, only to discover that an easy-to-fix mechanical vacuum leak was the source of the problem. Even worse, it might be a mechanical problem with the engine, such as not-so-easy-to-repair bent valves or a leaking head gasket.

Consequently, potential problems with the basic mechanical operation need to be separated from computer-controlled engine management system components, or

Despite the presence of an OBD-II diagnostic system, all the high-tech electronics associated with this not-so-glamorous 1996 General Motors 3100 V-6 engine are, for the most part, unable to determine if the engine is in good mechanical condition. If the basic engine has serious problems, emissions can reach levels that will cause a PCM to set DTCs. Vacuum leaks, low compression, coolant leaks, and worn-out parts can all lead even experienced technicians down the wrong diagnostic path in pursuit of false trouble codes. Performing basic diagnostic engine testing helps separate OBD-II system problems and emissions-related electronic problems from more basic mechanical engine failures.

OBD-II-related electronic system components, and investigated as a cause of a particular problem. This should preferably happen before commencing a lengthy investigation into DTCs that have been set that identify electronic components as a source of drivability issues. So with that caveat in mind, let's take a look at basic automobile engine operation and some simple tests that can be performed to quickly determine if an engine is mechanically sound. So, put on a thinking cap, and get ready to do a little automotive detective work.

Whether an engine has 4, 6, 8, or even 10 or 12 cylinders, the manner in which it creates power has not changed since the invention of the four-stroke engine in 1876. Nikolaus Otto, Gottlieb Daimler, and Wilhelm Maybach, all from Germany, were the first persons to discover and design a four-stroke method of burning gasoline in a piston-powered internal combustion engine. In 1886, Daimler went on to invent the world's first four-wheel vehicle to use an internal combustion engine. This vehicle was dubbed the automobile. In fact, the word "automobile" is derived from Nikolaus Otto's last name, and the way in which this earliest four-stroke internal combustion engine operated was often called an auto (Otto) cycle.

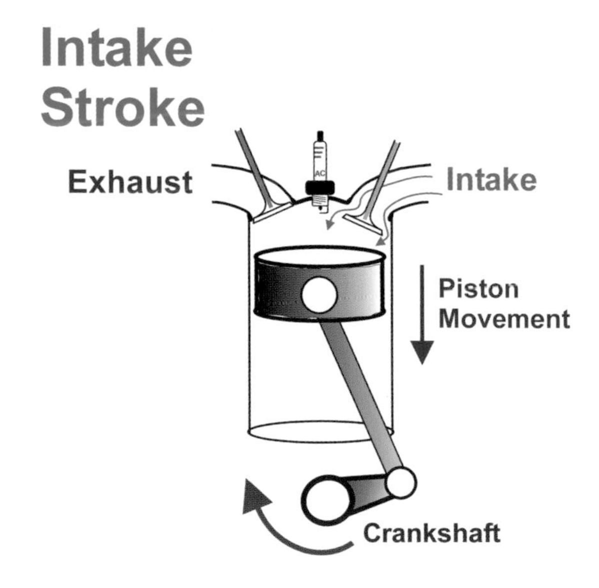

Fig 6-1—The first stroke of a four-stroke auto (Otto) cycle is the intake stroke. As the piston moves downward inside the cylinder, the intake valve is opened and air mixed with fuel is pushed into the cylinder by atmospheric pressure, since a low-pressure area was created by the rapidly moving piston. The intake stroke fills the cylinder with air and fuel, which will later be burned.

Four-Stroke Cycle

Understanding how an engine's four-stroke cycle works is a good starting point for understanding what to look for in an engine that's running as it's supposed to. Engine testing is relatively straightforward, and can be accomplished with simple tools such as vacuum gauges or compression testers. However, vacuum gauge readings and compression testing are somewhat meaningless, unless one understands what's taking place inside a running engine.

All four-stroke engines are, in reality, air pumps. The four-stroke cycle is the mechanical means that is used to pump air in and out of the engine. A four-stroke cycle is best conceptualized using an example of a single-cylinder engine, like those found in small motorcycles or lawnmowers. During a four-stroke cycle, a piston (connected to the crankshaft via a connecting rod) moves up and down inside a cylinder as the engine's crankshaft rotates. Each of these piston movements is considered a stroke—whether it be in either an up or down direction. A four-stroke cycle begins with the piston at the top of the cylinder. There are two valves—an intake valve and an exhaust valve—that are operated by the engine's camshaft, which rotates at one-half the speed of the crankshaft. The intake valve starts to open to allow fresh air and fuel to be drawn into the cylinder at the same moment the piston starts moving downward. This phenomenon is known as the *intake* stroke (see Figure 6-1). The camshaft then times the sequencing of the alternating opening and closing of the intake and exhaust valves to allow the four-stroke cycle to operate. This is known as camshaft timing and it simply means that the valves will open and close in the correct timing in relationship to the movement of the piston and crankshaft. A spark plug then ignites the air and fuel mixture at exactly the correct moment in time during the four-stroke cycle to burn the air and fuel inside the combustion chamber.

Once the piston is at the bottom of the intake stroke, the intake valve starts to close. The piston then starts to move upward and the intake valve closes completely. This is the beginning of the *compression* stroke, the second of the four strokes. The compression stroke squeezes or compresses the air/fuel mixture in order to make it more volatile so it can be more easily ignited by a spark plug. The amount the air/fuel mixture is compressed is referred to as an engine's compression ratio. Compression ratios are expressed as a ratio representing the number of times the air/fuel mixture is actually compressed from its original volume. For example, if the compression ratio for a particular engine is 11:1 that means that the mixture of air and fuel has been compressed 11 times its original volume by the time the piston reaches the top of the compression stroke (see Figure 6-2).

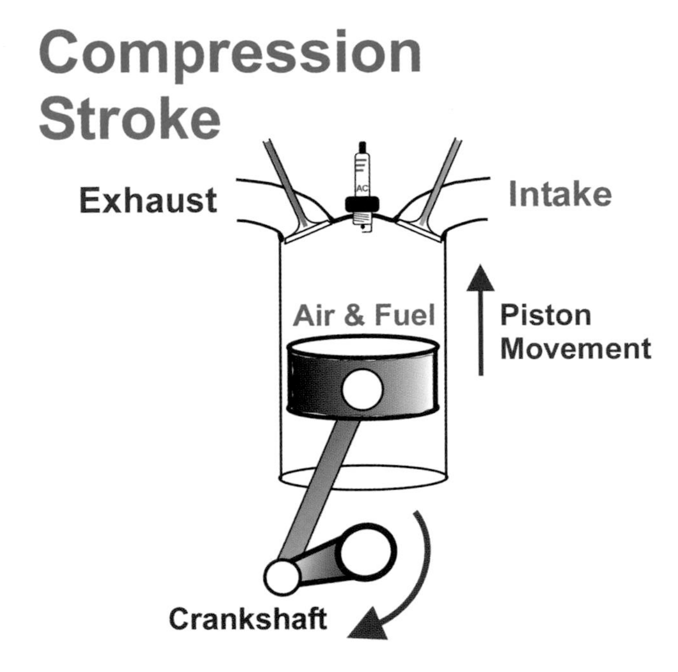

Fig-6-2— What is known as a *compression* stroke follows directly after completion of the *intake* stroke in a four-stroke internal combustion engine. As the piston begins traveling upward in the cylinder, both the intake and the exhaust valves are closed during the *compression* stroke. The piston compresses the air/fuel mixture as it pushes it upward. Compressing the air/fuel mixture makes it more volatile or easier to burn. An air/fuel mixture will not simply ignite from just a spark from a spark plug if the mixture is not compressed first. The air/fuel mixture is compressed to nearly 10 times smaller than the volume of the cylinder in a 10:1 compression ratio engine.

In the moment before the piston reaches the top of an engine's compression stroke, the ignition system sends a high-voltage spark to the spark plug, causing it to fire and ignite the air/fuel mixture. This sequence is the start of the *power* stroke, the third stroke in a four-stroke cycle (Figure 6-3, p. 122). The air/fuel mixture does not explode, but rather burns at an even, controlled rate. The combustion of burning gases causes a rapid increase in heat to over 5,000 degrees Fahrenheit inside the combustion chamber, which, in turn, causes the pressure inside the cylinder to build to several tons per square inch, all of which pushes down on top of the piston. Since the piston is connected to the crankshaft via a connecting rod, the crankshaft rotates as a result of the piston exerting downward force as it moves down the cylinder. The twisting action of the crankshaft in turn results in power or torque being transmitting to the engine's flywheel. The flywheel is connected to whatever the engine was designed to power—a car's transmission, motorcycle, lawnmower, and so on.

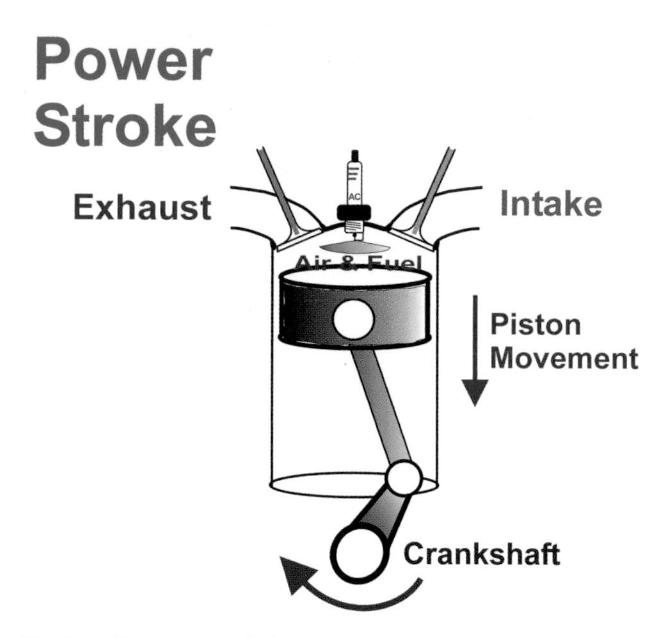

Fig-6-3—The *power* stroke is the third in the sequence of piston strokes in a four-stroke cycle. Immediately before the piston reaches the top of its power stroke, the ignition system sends a high voltage spark to the spark plug. When the spark jumps the spark plug's air gap, the combustion process is ignited or started. Heat from the burning air/fuel mixture then causes a sharp increase in cylinder pressure, which in turn places several tons of pressure on the top of the piston. The power stroke of a four-stroke cycle is the only stroke that produces work—or power, energy, or heat, however you want to conceptualize it) The other three strokes, while essential to the auto cycle process, simply use up energy instead of producing it.

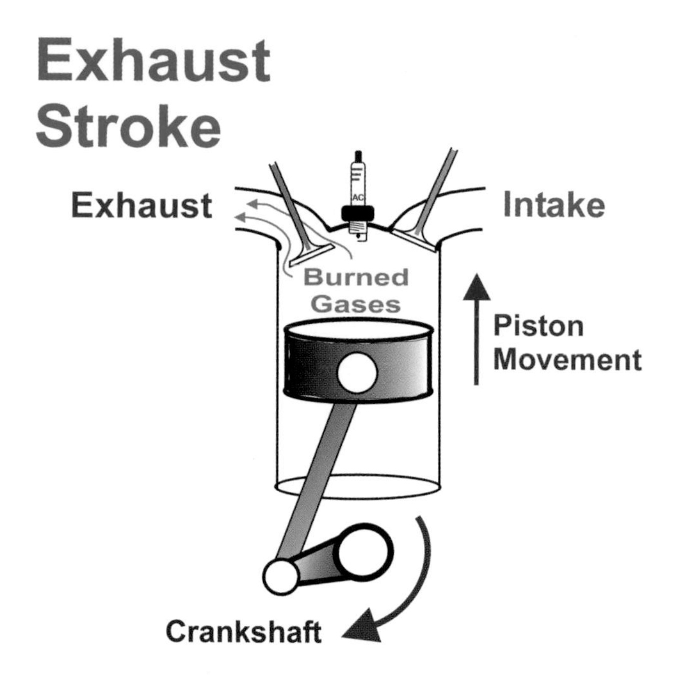

Fig-6-4—The *exhaust* stroke is the last stroke that occurs in a four-stroke internal combustion engine. Once a piston reaches the bottom of the third or power stroke, the exhaust valve opens and the piston starts to move upward. The upward movement of the piston pushes the exhaust gases out of the cylinder. Just before the piston reaches the top of the exhaust stroke, the intake valve starts to open and the piston reverses its upward direction of travel, so that it is now traveling downward, and the four-stroke cycle starts all over again with the intake stroke.

As the piston reaches the bottom of its power stroke, the exhaust valve starts to open, allowing the burned gases to escape from the cylinder. As the piston moves upward on the *exhaust* stroke, the exhaust valve opens all the way and the cylinder is cleared of exhaust gases, which is the end of a four-stroke cycle in an internal combustion engine (see Figure 6-4.) Then, just before the piston reaches the top of the exhaust stroke, the intake valve starts to open and the four-stroke cycle starts to repeat all over again. The four piston strokes of a four-stroke Otto cycle—*intake*, *compression*, *power*, and *exhaust*—are frequently known in automotive jargon by the much more colorful descriptive terminology of suck, squeeze, bang, and blow.

A four-stroke cycle takes two complete engine revolutions to complete—the intake and compression strokes comprise one full engine revolution, while the power and exhaust strokes complete another revolution. However, a four-stroke cycle occurs many times per second when an engine is running at high speeds. In fact, a single-cylinder engine turning 5,000 revolutions every minute (this is called engine rpm) will complete a full four-stroke cycle 41.6 times every single second. Engines with multiple cylinders work in the same manner. Since all of the cylinders are connected to a common crankshaft, each of the cylinders repeat the same four-stroke cycle every two engine revolutions—regardless of the number of cylinders an engine has.

Engine Vacuum

Although it's difficult to look under the hood of a vehicle at an engine while it's running, in order to visualize a four-stroke cycle occurring, the results of this process can be viewed with some simple, inexpensive tools. One of the most basic tools for this purpose is a vacuum gauge, a tool that has been around as long as the internal combustion engine. These tools are available at most auto parts stores for less than $20. A quick test with a vacuum gauge will provide enough information about overall engine condition,

Pictured is a compound vacuum/pressure gauge. The numbers on the inside scale (red and green ranges) represent inches of mercury, abbreviated as In-Hg; this scale is used for taking vacuum readings, and ranges from 0 to 30 inches. The numbers on the outside of the gauge (white numbers) is a scale representing centimeters of mercury, abbreviated as Cm-Hg. On the gauge, notice the white 0 range (where the needle is pointing) is large because 0 may vary slightly, depending on the altitude the gauge is being used at, due to differences in barometric pressure. Numbered scales at the top of the gauge are for testing fuel pump pressure on carbureted vehicles.

and how that condition may relate to any OBD-II trouble codes that may be present in the PCM's memory. It's a good idea for anyone who has never used a vacuum gauge before to practice the test procedures outlined in this chapter before spending any length of time trying to determine what's wrong with a "problem" engine. All automotive technicians, whether professional or home mechanics, must be able to identify and understand what the test results will be for an engine with no mechanical problems *before* beginning testing on an engine with something amiss. Consequently, being able to interpret vacuum gauge readings requires a fundamental understanding of how engines create vacuum. Thus, engine vacuum is an important concept to master before tackling any significant testing on a malfunctioning engine.

Contrary to popular belief, the air/fuel mixture consumed in the combustion process is not sucked into an engine, but is instead pushed by atmospheric pressure. Here's an illustration of how this process works: Suppose a person is standing on a beach looking out at the ocean. That person has about 40 miles of atmosphere above him/her. The weight of the atmosphere pushing down on the earth creates atmospheric pressure, which is measured in pounds per square inch. Atmospheric pressure at sea level is 14.7 pounds per square inch pushing in all directions (Figure 6-5, p. 124).

During an engine cylinder's intake stroke, the intake valve opens as the piston is moving down the cylinder. The movement of the piston increases the cylinder's volume faster than the opening of the intake valve allows atmospheric pressure to fill the cylinder; this process creates a vacuum (negative air pressure) inside the cylinder during the intake stroke. A vacuum is created whenever pressure inside the cylinder is lower than the surrounding atmospheric pressure. Air is pushed into the cylinder because relatively high atmospheric pressure flows into the lower negative air pressure inside the cylinder. How much the cylinder fills with atmospheric pressure is called volumetric efficiency (VE), and on most engines it's around 85 percent of total cylinder filling, or more. VE almost never reaches 100

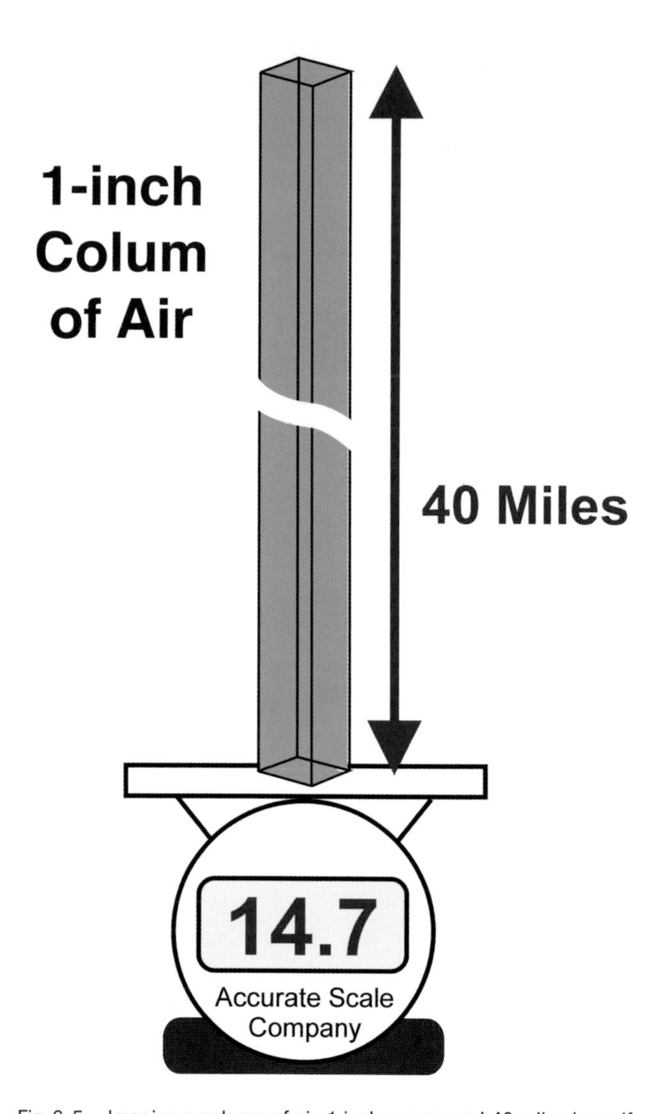

Fig-6-5—Imagine a column of air, 1 inch square and 40 miles long. If a scale could be used to weigh the air in the column, it would weigh 14.7 pounds at sea level—the weight of the air at sea level is also known as barometric pressure. Since pressure from the weight of the atmosphere (air) pushes in all directions, barometric pressure is 14.7 pounds per square inch at sea level. If the scale was relocated to 10,000 feet above sea level, atmospheric pressure would drop because there is actually less air present above the scale since air is less dense at higher altitudes; thus, barometric pressure is lower. It's often confusing, and merely coincidental, that barometric pressure of 14.7 pounds per square inch is a similar number to the ideal air/fuel ratio of 14.7:1 (14.7 parts of air to one part of fuel).

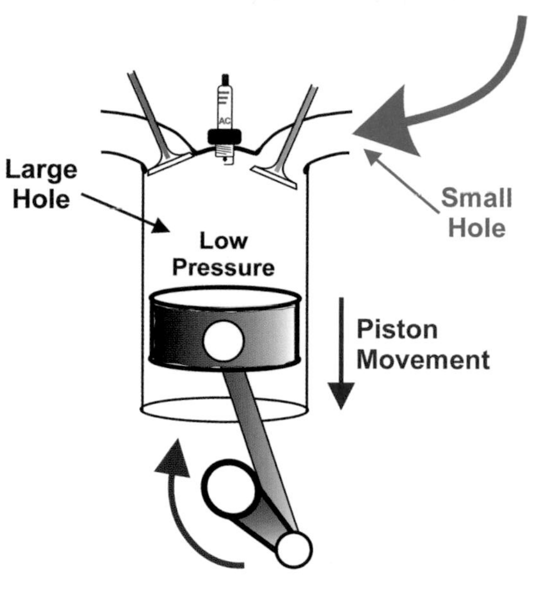

Fig-6-6—One of the laws of physics basically states that an area of high pressure will always seek out an area of low pressure and flow into it. In the cylinder in Figure 6-6, the piston is moving rapidly downward, thereby increasing the volume, or air space, of the cylinder above the piston. As a result of the relative size of the cylinder's volume compared with the diameter of the intake valve's opening, the incoming air cannot fill the cylinder with air fast enough to completely fill it so a low-pressure area is created inside the cylinder. The higher pressure of the outside atmospheric air is literally pushed down into the intake manifold, past the intake valve opening, and into the cylinder, with the atmospheric pressure basically trying to fill up the cylinder.

percent because the intake valve restricts the amount of air that can fill the cylinder. Engines equipped with turbochargers overcome this limitation on cylinder filling and are able to achieve greater VE because the turbocharger's air compressor pressurizes incoming intake air above atmospheric pressure, so the cylinder can fill to more than 100 percent of its volume. Depending upon the boost settings of a particular turbocharger, volumetric efficiency can actually top upward of 120 percent. Engines with two intake valves (as compared to one intake valve) and tuned intake runners can also fill an engine's cylinders close to 100 percent, since they also allow more air to flow into the cylinder at higher engine rpm.

Since air has to travel through a common intake manifold to get to the cylinders on a multicylinder engine, a constant vacuum is created inside the intake manifold. Thus, a vacuum gauge can be connected to the intake manifold to measure the pressure of the engine vacuum and to otherwise perform vacuum testing. Because vacuum readings are directly related to how well an engine seals against atmospheric pressure, taking vacuum readings at different engine operating conditions will provide an overall picture

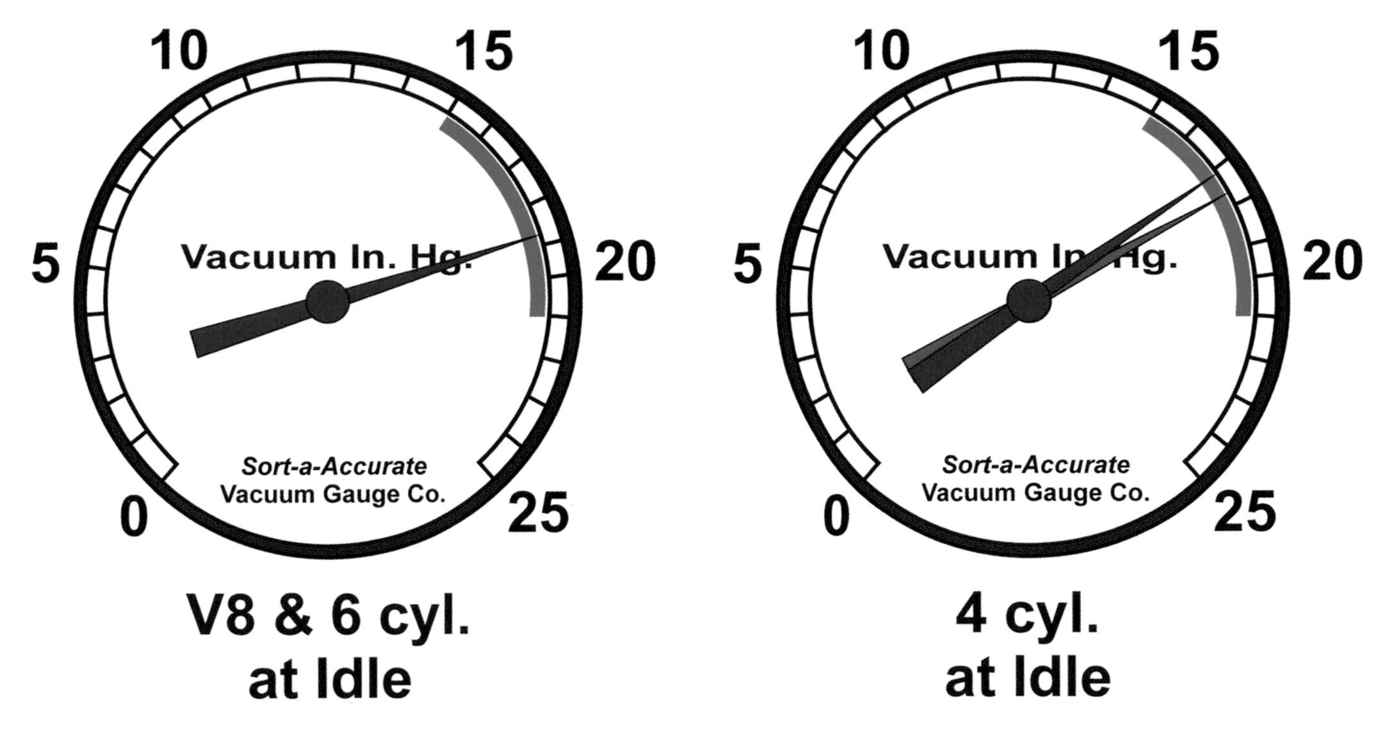

Fig-6-7— An engine idle vacuum test confirms whether an engine is in good mechanical condition, or free from any vacuum leaks. At idle, an engine in good condition and with no vacuum leaks should produce a gauge reading on a scale of between 15 and 21 In-Hg of vacuum. When connected to a V-8, V-6, or inline-six-cylinder engine, the needle on a gauge should hold steady. On certain four-cylinder engines, the needle may pulse slightly, which merely represents the actions of individual cylinders on their intake strokes.

of engine mechanical health, thus saving needless hours of scan tool/OBD-II testing and reading trouble codes only to subsequently discover the engine has a leaky intake manifold, has bent valve(s), or is simply worn out from abuse, high mileage, or lack of maintenance.

Engine Idle Vacuum Testing

To perform vacuum testing with the engine at idling speeds, a vacuum gauge should always be connected directly to the intake manifold or tapped into a vacuum line with a T fitting. Avoid vacuum lines that connect to canister or exhaust gas recirculation (EGR) valves, as they may carry vacuum signals lower than those produced by engine vacuum, and vacuum test results may be misleading. Also, try to avoid vacuum lines that contain vacuum control valves or restrictors for the same reasons. With the engine's speed maintained at idle and at normal operating temperature, a vacuum gauge needle should, more or less, register steady (depending on the number of cylinders an engine that is being tested has). Vacuum readings on some four-cylinder engines may show a slight needle pulse due to camshaft timing that causes individual cylinder's intake strokes to pulse the vacuum gauge needle. These types of vacuum readings can take place on engines with high-performance camshafts, even those with six or eight cylinders as well.

Vacuum gauge readings that are between 15 and 21 inches of mercury (In-Hg) with a steady needle are in the normal range and indicate an engine that is sealing properly. The more cylinders present on an engine, the less the needle will pulse at idle because there are more intake strokes per engine revolution. For this reason, normal needle movement for a V-8 is almost nonexistent; six-cylinder engines may produce very slight pulsing. The significance of a steady vacuum gauge needle is that it confirms that all of the engine's cylinders are operating in the same manner, so the pressure drop for each cylinder within the intake manifold will be the same (causing a steady needle gauge reading); this confirms that all cylinders are sealing properly, and also, none of the valves, piston rings, or head gasket(s) are leaking (see Figure 6-7).

If a vacuum gauge needle is not steady, but is wildly bouncing back and forth, this needle action suggests that a serious problem probably exists within an engine. A pulsing needle could mean that one or more cylinder(s) is not contributing to the overall level of intake manifold vacuum to

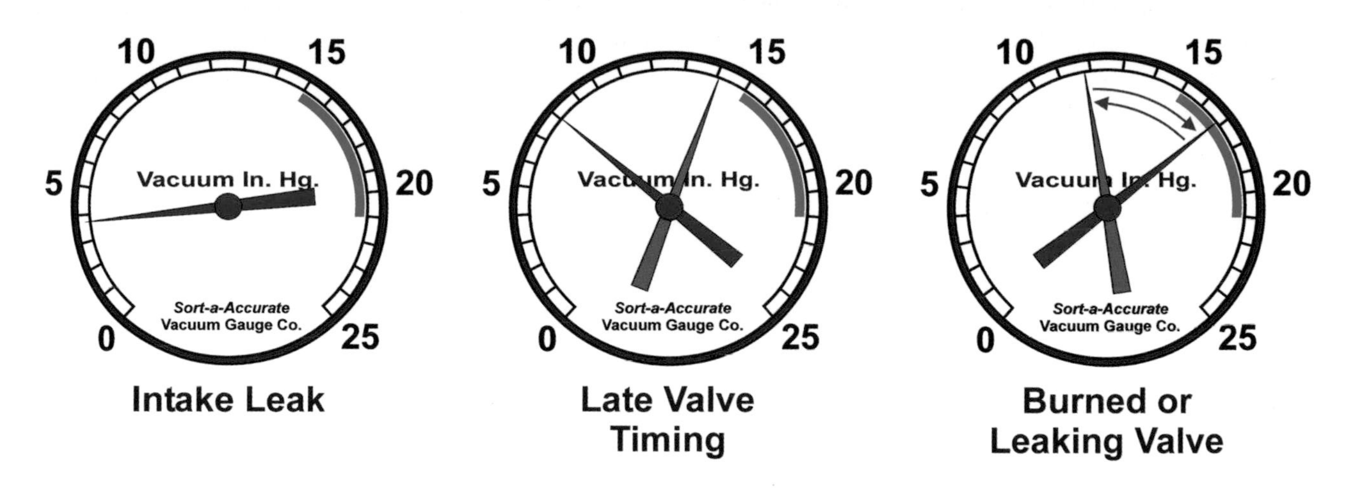

Fig-6-8—All vacuum gauges show various engine conditions at idle. The vacuum gauge shown at the far left of Figure 6-8 has a steady needle, but a low reading, indicating that an intake manifold gasket or vacuum fitting is leaking, and thereby causing a lowering of engine vacuum. The needle on the center gauge is floating between 8 and 14 In-Hg (inches mercury), and establishes what engine vacuum looks like on an engine with late valve timing caused by a slipped timing belt. The vacuum gauge on the far right of the figure reveals a steadily dropping, or pulsing, needle thus confirming a burned or leaking valve, or head gasket problem.

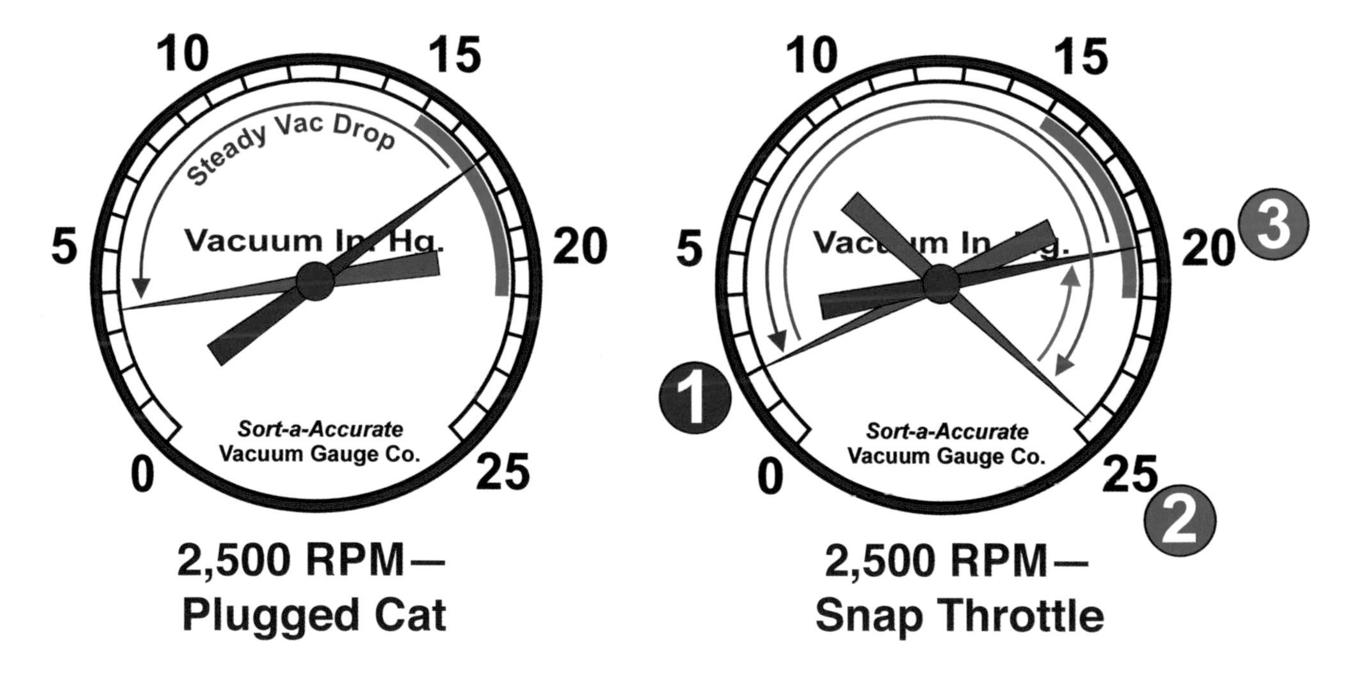

Fig-6-9—The vacuum gauge on the left in Figure 6-9 shows the levels of engine vacuum that are created by a plugged catalytic converter or other exhaust restriction. If engine speed is held steady at 2,500 rpm, but the vacuum needle slowly drops, a plugged catalytic converter or other exhaust restriction is definitely indicated as a problem. This takes place because the exhaust restriction prevents exhaust gases from exiting the cylinder and raises cylinder pressure higher than normal, thus creating a high pressure, or low vacuum reading. The vacuum gauge on the right shows a healthy engine when the throttle is snapped open suddenly. On a healthy engine, when engine speed is held steady at 2,500 rpm, the needle is close to 20 In-Hg, but when the throttle is snapped open suddenly, the needle drops from 20 to 2 In-Hg (position 1 on the right hand gauge. As soon as the throttle is let go via the throttle return spring closing it, the needle quickly rises to 25 In-Hg (position 2 in blue), but then finally returns back to the normal range of idle vacuum at about 20 In-Hg (position 3 in green).

the same extent as all the other cylinders, and is therefore not doing its job properly or completely. In such an instance, further vacuum testing is required. If engine idle vacuum is steady, but lower than what would typically be expected for a normal range of 15 to 21 In-Hg, there could be several causes for this. The most common and likely source of such a reading is an intake manifold vacuum leak. The presence of a vacuum leak in an intake manifold could be a result of a poorly sealing gasket, or could simply be a loose or cracked vacuum hose. A vacuum leak would cause the pressure inside an intake manifold to rise closer to atmospheric pressure, thus lowering the vacuum gauge reading. (Remember, the vacuum gauge will read 0 In-Hg with no vacuum present.)

Another cause of low vacuum readings results from ignition timing that may be retarded. However, retarded ignition timing is not usually a problem on OBD-II-equipped vehicles, since ignition timing is not adjustable. (On older vehicles with ignition distributors, this is a possibility that should be checked out.) However, retarded camshaft or valve timing can be a problem on OBD-II vehicles and is typically caused by a timing belt that is worn out, causing it to jump one or several teeth (resulting in low vacuum readings). If a timing belt has slipped a tooth or more, due to age and wear, the valves become out of sync with an engine's four-stroke cycle, which invariably causes a low vacuum reading. On some engines, whenever a timing belt slips a tooth, the valves can come into contact with the tops of the pistons, resulting in bent valves that fail to seal properly. If this happens, and either the intake or the exhaust valve(s) do not seal properly, their respective cylinders will not be able to create the same vacuum pressure level the other properly functioning cylinders produce. As a result, the unmatched or disproportionate pressure levels in all of the cylinders cause the vacuum gauge needle to swing wildly back and forth at idle speeds (see Figure 6-8).

Testing Engine Vacuum at 2,500 RPM

Performing vacuum testing while holding the engine steady at 2,500 rpm can reveal several different engine problems. Engine vacuum readings at 2,500 rpm should be steady, regardless of the number of cylinders on an engine, provided there is nothing wrong with the engine. If the needle on a vacuum gauge bounces or pulses, then one or more cylinders are not sealing properly, which could be attributable to either sticky, burned, or bent intake and/or exhaust valves or a blown head gasket.

To test for worn piston rings or cylinders, or to simply examine overall engine condition, hold engine speed at a steady 2,500 rpm for 15 seconds while a vacuum test gauge is attached to the engine's intake manifold. After 15 seconds, snap the throttle all the way open and then close it. (Simply letting go of the throttle will close it.) When the throttle is held at 2,500 rpm, the test gauge needle should be steady; however, once the throttle is snapped open, the test gauge needle should instantly drop to 2 or even 0 In-Hg After the throttle is let go, the test gauge needle should return to a high vacuum reading of around 25 In-Hg and then taper off to the normal idle vacuum range of about 20 In-Hg. Typical needle readings such as these indicate a sound mechanical engine. However, if snapping open the throttle does not lower the vacuum reading to 2 In-Hg or less, a restricted intake somewhere (usually a plugged air filter) is indicated. Conversely, if letting go of the open throttle doesn't increase engine vacuum by a range of at least 2 In-Hg above the reading obtained when engine speed was held steady at 2,500 rpm, then worn out piston rings or cylinders are most likely a problem (see Figure 6-9).

A good test for a plugged exhaust caused by either a vehicle being driven over a curb and denting the exhaust, or by a plugged-up catalytic converter can accomplished by holding engine speed steady at 2,500 rpm for 15 seconds, and then checking the vacuum gauge needle to see if it slowly drops (the vacuum gauge is connected to the intake manifold). A plugged catalytic converter will cause excessive exhaust backpressure to build up. If excessive exhaust backpressure exists, the vacuum gauge needle will slowly drop to a level of 5 In-Hg or less. A plugged catalytic converter is the most common reason for this type of low vacuum reading at steady rpm of 2,500 (see Figure 6-9).

A common misconception about ignition misfires is that they will cause a vacuum gauge's test needle to bounce wildly. This phenomenon is simply not true, since a vacuum gauge only measures an engine's ability to properly go through the four-stroke cycle and act as an air pump, as stated earlier. Consequently, if a cylinder has an ignition malfunction that results in an ignition misfire, the engine is still pumping air and fuel in and out of its cylinders, but it is no longer burning the air/fuel mixture in the specific misfiring cylinder. As long as the misfiring cylinder is still sealing during the four-stroke cycle, the presence of a misfiring cylinder will not greatly affect vacuum gauge needle movement. In fact, the only reason a vacuum gauge needle moves at all because of a cylinder misfire would be as a result of an engine slowing down every time the specific misfiring cylinder fails to properly complete its power stroke.

Cranking Engine Vacuum Testing

The main reason for performing cranking engine vacuum tests is that it is a good method of determining the reason

The compression tester in the photo comes complete with two adapter hoses for use with different spark plug thread sizes. The compression gauge pictured has a quick disconnect attachment for the hoses. This gauge is capable of holding pressure readings until after the engine has stopped cranking, making it easier to record them. The Schrader valve shown in the picture on the side of the fitting is the exact same valve commonly used to fill tires with air. The purpose of a Schrader valve is to release the pressure built up by engine compression, thus returning the test gauge needle to zero after completion of a compression test.

A more accurate method for finding vacuum leaks involves the use of propane gas. Take a propane bottle and connect a small section of hose to it, and then, with the engine running, crack open the valve on the propane bottle while directing the free end of the hose all over and around all vacuum hoses and fittings, as well as the intake manifold. Any vacuum leaks present will draw in the propane gas and cause a momentary change in engine rpm. An even better method would be to perform the same test but also use a scan tool connected to the vehicle. Then watch the front upstream O_2 sensor's output voltage for instant voltage changes of up to 0.9 volts or higher. When the propane is sucked into the vacuum leak, the oxygen in the exhaust gas is reduced, causing the high O_2 sensor voltage reading. Remember that there is a time lag between when the O_2 sensors picks up the additional fuel and when the scanner displays the high-voltage reading.

CAUTION: Now, in an effort to keep all lawyers happy, bear in mind that propane is a highly flammable gas that can cause severe burns and become explosive if used near heat in a nonventilated space. Consequently, always use caution when performing any of these tests that involve the use of propane. In addition, always use common sense when handling propane. For example, if you can hear a loose or disconnected spark plug wire arcing to ground (it makes a snap, snap, snap sound) DO NOT use propane to search for any vacuum leaks until the plug wire is repaired, since doing so could cause a small fire or explosion, which could obviously result in damage to a vehicle and potentially catastrophic injury or injuries to people. Thus, when using any type of fuel, also use some common sense to keep you safe. Remember, if you manage to blow yourself up, or your garage or vehicle, when using this leak detection method, don't say you weren't adequately forewarned!

Compression Testing

If, during the performance of any engine vacuum test, the vacuum gauge needle starts to bounce wildly back and forth, there is a strong chance that one or more of the cylinders is not sealing or closing completely or has other mechanical problems. Vacuum testing is not sophisticated enough to indicate which cylinder has the problem or is causing a particular problem, so other additional methods of testing are required. Of course, this assumption begs the question of why it matters which cylinder is having a sealing problem if the engine has to be rebuilt anyway. While this sentiment is true to some extent, having more knowledge and information about why, and specifically where, the engine is failing to make a complete seal may help with the determination of whether a complete engine rebuild is necessary or not, or whether a simpler valve job, or even simpler head gasket replacement might just as well do the trick. Of course, there are always those of us who just have to know which cylinder is failing to seal properly just so we can satisfy our compulsive mechanical tendencies.

Compression testing is a good way of identifying and confirming which cylinder is having a mechanical problem. Compression testers are relatively inexpensive, and are available at most auto parts stores for around $40. More expensive compression gauges are naturally more accurate, and also come complete with more adapters and accessories than standard models. Handheld compression testers designed to be held inside a spark plug hole are difficult to use and frequently don't produce consistent results. Compression testers fitted with hoses that screw into a spark plug hole are much easier to use and have better predictability in the accuracy department.

The first step in performing compression testing is to get the engine hot. This allows all of the internal engine parts to expand to their normal size, and to seal as well as they possibly can during the test. The next step is to remove all the spark plugs. On some vehicles this can be quite a project, sometimes even taking hours just to gain access to them. The fuel injection and ignition systems must also be disabled. Usually the PCM fuse(s) or ignition fuse(s) can simply be removed to prevent these systems from operating. It's not a good idea nor a safe practice to

A remote starter switch makes compression testing easier to accomplish because the compression gauge can be viewed at the same time the starter is operating, requiring only one technician to perform the test. The remote starter switch in the photo was self-made from a plastic 35-millimeter film container, a heavy-duty switch from Radio Shack, some flexible wire, and two alligator clips. Remote starter switches can also be purchased at auto parts stores and online.

This compression gauge has a range from 0 to 300 psi (the outside number scale on the gauge). The inside number scale is in kPa x 100, or kilopascals for those technicians who are metrically inclined. The quick disconnect fitting at the bottom of the gauge allows various hoses with spark plug thread adaptors attached to the gauge. The Schrader valve on the right side of the gauge's quick-disconnect fitting is used to release gauge pressure after the completion of each cylinder's compression test.

have fuel injectors spraying fuel with open spark plugs removed and ignition coils or cables disconnected, since doing so could create a high-voltage spark in an enclosed space (possibly leading to a fire or other damage). Start the test by screwing a compression tester's hose into the first (No. 1) cylinder's spark plug hole. Next hold the engine's throttle all the way open. You may want to temporarily block the throttle open, because it must remain so during the entire compression test sequence. The open throttle allows the maximum amount of air to enter the engine during testing.

The next step involves cranking the engine over. Have a friend inside the vehicle to operate the starter motor via the ignition key or a remote starter switch. The best way to perform a compression test is to connect a remote starter to a starter solenoid, since in this way, a sole technician can control the starter motor while simultaneously watching the test gauge. To start the test, the engine should be cranked over long enough so that the cylinder being tested goes through four complete compression strokes, which can easily be determined by watching the compression gauge needle as it pulses. Each pulse represents one compression stroke. The numerical scale on most compression gauges represents pounds per square inch (psi). While the test gauge needle is being watched, compression numbers for the first and fourth strokes should be recorded. Repeat the test for all other cylinders. Needless to say, a vehicle's battery should be in good operating condition during compression testing, since it will have to crank the engine over a number of times. If the battery is weak, the last cylinders tested could possibly have a lower range of numbers displayed, due to a slower engine cranking speed as the battery loses energy.

So, what exactly are the good and bad ranges of numbers that can be expected from compression testing? Some manufacturers have published cylinder compression numbers in their service manuals, and occasionally will also list minimum compression numbers and permissible variations between cylinders. As a general rule, the lowest compression test readings should still be within 75 percent of the highest reading possible for all the cylinders on the engine. For example, if the highest cylinder compression test reading yields 140 psi, the lowest reading should not be less than 105 psi (or 75 percent of 140). Another general rule to remember is that if compression pressure is below the range of 85 to 100 psi in any one cylinder, that cylinder is worn out, and the engine, unfortunately, needs to be rebuilt.

Here are some guidelines for interpreting compression test results:

If compression increases steadily after all four compression strokes (remember, the cylinder should go through four complete four-stroke cycles) of the four-stroke cycle and is still within manufacturer's recommended specifications, or within 75 percent of the highest reading obtained for any one cylinder (when compared to all cylinder readings), then the cylinder being tested is good.

Battery Open Circuit Voltage

State of Charge	Battery Voltage
100%	12.65 or more
75% - 100%	12.45v
50% - 75%	12.24v
25% - 50%	12.06v
0% - 25%	11.89 or less

Fig-6-11—Measuring open circuit voltage is an accurate method for determining a battery's state of charge. Before performing any testing of this nature, make sure all electrical loads are off before proceeding further. If a vehicle's battery has recently been charged, the surface charge will also have to be removed prior to any battery testing; this can be accomplished simply by disabling the ignition or fuel system, and then cranking the engine over for about 15 seconds. Then let the battery rest for five minutes before testing for open circuit battery voltage.

If compression is low on the first compression stroke, and then gradually increases during the remaining three compression strokes, but still fails to reach the manufacturer's good or minimum permissible range of compression pressure, the piston rings and/or that specific cylinder being tested is/are probably worn out.

If compression is low on the first compression stroke, and only increases slightly during all of the remaining compression strokes, either the valves may be sticking or burned, or the piston rings may be broken.

If compression is simultaneously low in two adjacent cylinders, suspect a blown head gasket, as it is likely leaking between both cylinders.

If a compression test gauge reading ever yields a higher range of numbers than what is recommended in manufacturer's specifications, that cylinder(s) may have excessive carbon deposits in its combustion chamber, which are causing higher than desirable compression readings.

If it's determined that compression is low in any cylinder due to an inadequate piston ring sealing problem, a wet compression test can be performed, which will provide significant information about the condition of the piston rings. To perform a wet compression test, squirt a small amount (about one tablespoon) of engine oil into the specific cylinder being tested and then crank the engine over a few times to let the oil soak in. Then, repeat the compression test. If the compression number increases by 5 percent (usually about 15 psi), that cylinder probably has worn piston rings. However, if compression doesn't increase at all, then the sealing problem is most likely in the cylinder head's valves or the head gasket seal. Bear in mind, a wet compression test won't work on engines with a horizontally opposed design, like those found on Volkswagen Bugs, air-cooled Porsches, and some Subaru models. Because cylinders are positioned on their sides in these vehicles, the oil added into the cylinder for a wet compression test won't

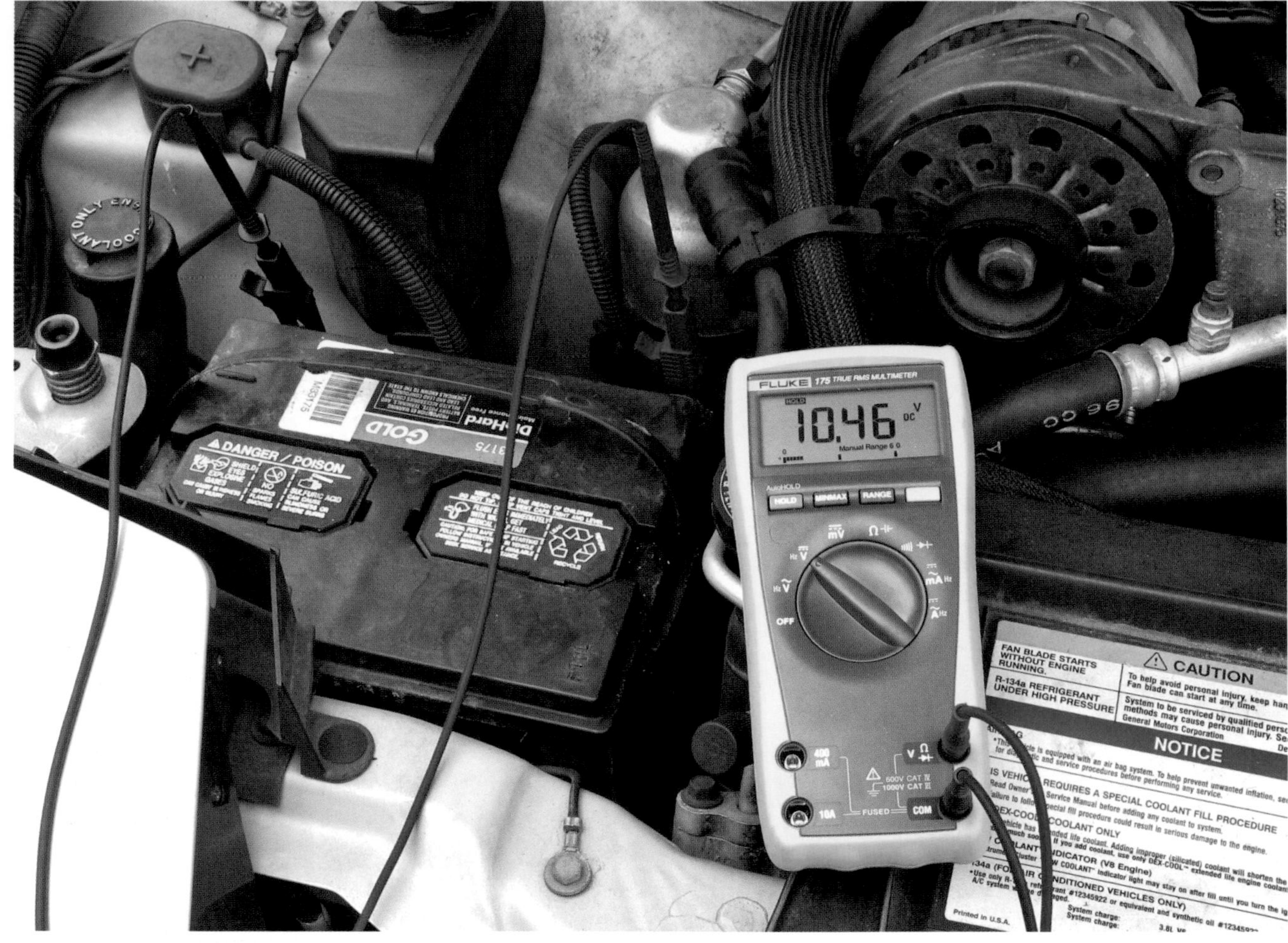

Electrically loading the battery by cranking the starter motor over is a good method to test for battery starting capacity. With a digital voltmeter connected directly to the battery, crank the engine over for about 15 seconds. Record the voltage displayed just before the starter has stopped cranking the engine. If the voltage reading is 9.6 volts at 79 degrees Fahrenheit or higher, the battery is in good condition. The battery in the photo is showing 10.46 volts, which indicates it is in like-new condition.

be able to work its way completely around the piston rings, so they won't seal any better than they would when a dry compression test is performed.

Battery and Electrical System Testing

Because all OBD-II systems are electrical in nature, it makes sense that the vehicle's battery and charging system should be operating properly before commencing any OBD-II testing. The PCM performs hundreds of electrical measurements that compare sensor signals to each other and to references voltage within the PCM itself. If the vehicle has battery- or charging-related problems, the PCM could set numerous false DTCs caused by electrical system problems. This is especially true if the vehicle being tested is experiencing a number of different DTCs that are all related to sensor signals or PCM output failures, since these could actually all be related to a bad battery or poorly operating charging system instead.

Now is an appropriate time to go over some basic battery and charging system tests. For more detailed test procedures on automotive electrical systems, please read the book *How To Diagnose and Repair Automotive Electrical Systems* by Tracy Martin, published by Motorbooks, as it will provide easy-to-understand procedures and explanations on the subject. This related automotive electrical book is available at the author's Internet website at www.tracyAmartin.com.

There are several methods available for testing a battery, ranging from using expensive, professional-grade test equipment to using only a digital voltmeter in conjunction with an engine's starter motor. Regardless of which test is used, the vehicle's battery should be fully charged. Simply using any digital voltmeter, a battery's state of charge can easily be determined. An *open circuit* voltage test can determine a car battery's state of charge, whether the battery is conventional (utilizing filler caps to add water) or maintenance-free (no filler caps).

The most basic and most frequently used test for determining a battery's state of charge is an open circuit voltage test. This test determines the relative state of charge of the battery for purposes of further testing. If a vehicle has recently been driven, its battery needs to rest for at least 10 minutes without any electrical load being placed on it before an open circuit voltage test can be performed. It is important to make sure all electrical loads are turned off before testing begins. To start the test, connect a digital voltmeter directly to the battery and then measure its voltage. See chart in Figure 6-11 to determine states of charge for any given voltage reading. If the battery's state of charge is less than 75 percent of its full capacity, the battery must be charged further before any additional battery or related OBD-II testing can occur. Battery chargers are available from most auto parts stores and Sears. After charging the battery, if voltage does not increase to 12.6 volts or higher, the battery should be replaced.

There is a second battery test that can indicate how much cranking capacity a battery has available to start an engine. This test stresses a battery by placing an electrical load on it. The fuel injection or ignition system will have to be disabled in order to perform this test. With one of these systems disabled, the engine can be cranked over without starting for the battery load test. Typically, just removing ignition or PCM power fuses will prevent the ignition and/or fuel systems from operating, as well as removing the fuel pump fuse. To perform this test, connect a voltmeter to the battery and crank the engine for about 15 seconds. Next, watch the voltmeter reading just before the starter stops cranking—a good battery should have a minimum loaded voltage of 9.6 volts at 70 degrees Fahrenheit. If testing is done in a cold climate (ambient temperature around 40 degrees Fahrenheit), minimum voltage should read 9.3 volts. If loaded voltage is less than the minimum, the battery is weak or tired and well on its way to being the source of an engine starting problem. If the battery is bad, loaded voltage will drop way off (to less than 7 volts) within the first few seconds of engine cranking. A weak, old, or tired battery simply will not pass this test and should be replaced.

After testing a vehicle's battery, its charging system should be checked. Again, a faulty charging system could cause the PCM to set false trouble codes. The following is a description of two charging system tests: Leave the voltmeter connected to the battery as it was during the open circuit voltage test, and replace whatever fuses were removed in order to perform the previous tests. Then start the engine. The battery should register charging system voltage of at least 0.5 volt more than was produced by the open circuit voltage test performed earlier. Appropriate vehicle service manuals will list correct ranges for charging voltages; they are usually between 13 and 14.5 volts, although naturally there are some exceptions. In any event, if charging voltage exceeds 14.5 volts, or falls below 13 volts, the charging system needs to be checked further for potential problems. Over- or undercharging voltage conditions could possibly cause a PCM to set charging system-related DTC(s), but don't count on it.

The final charging system test checks the condition of diodes inside the alternator. Diodes are solid-state one-way electrical valves, so if they start to leak AC voltage into a vehicle's DC voltage system, a PCM could set various false DTCs. To perform the last test, connect the red voltmeter lead from a digital voltmeter directly to the back of an alternator's output wire (usually the large wire on the back of the alternator). Next, set the scale on the voltmeter to read AC millivolts. AC voltage should not exceed .055 volt AC (55mV AC) with the engine running and several accessories turned on. However, if AC voltage is greater than 55mV, one or more diodes could be about to malfunction, so the alternator should be checked further, and possibly replaced.

Ignition Misfire Testing

So far we've discussed how to diagnose basic mechanical engine problems that relate to cylinder sealing and compression. Another common problem that has existed since the invention of the automobile—and is still around today—is ignition misfire. To put modern ignition misfires and their related DTCs into perspective, let's take a look at some early ignition systems. Before 1975, automotive ignition systems were high-maintenance, requiring attention every 10,000 miles or sooner. Keep in mind, 10,000 miles in 1975, and earlier, was a lot of driving mileage to rack up, as people tended to live closer to their work, and the interstate system of highways was not as extensive as it is today. Nevertheless, mechanical points, the condenser, distributor, spark plug wires and boots, and spark plugs all wore out or needed frequent adjustment.

With the introduction of more stringent emissions controls in the mid-1970s, especially those affecting the exhaust gas recirculation (EGR) system, spark plug energy had to be boosted considerably in order to reliably ignite lean air/fuel mixtures caused by EGR application at part throttle. Furthermore, the lean air/fuel mixtures that catalytic converters required in order to operate caused spark plug firing voltages to need to be increased even further. The addition of catalytic converters to vehicles resulted in the creation of a chemical environment inside an engine's combustion chamber that served to increase a spark plug's air-gap resistance well beyond the range of what older ignition systems could provide (only about 20,000 volts). In fact, spark plug firing voltages could reach 50,000 volts or more on these early emission systems. In addition to the mechanical necessity for higher voltage outputs, the federal government also mandated that ignition systems had to be maintenance-free for at least 50,000 miles.

Today, distributorless ignition systems equipped with coil packs and plug wires, as well as coil-over-plug systems that have no spark plug wires, can output voltages as high as 100,000 volts. These OBD-II-monitored ignition systems are almost maintenance-free, since spark plug replacement intervals are often recommended at no less than 100,000 miles, and there are no associated adjustments that have to be made either, regardless of how many miles are on the odometer.

Despite their high reliability, ignition systems on OBD-II vehicles can still misfire. In fact, there are over 90 generic ignition system misfire-related OBD-II DTCs in existence. These trouble codes bear DTC numbers that start with the prefix P03XX, and include, as examples, the following potential reference DTCs: individual cylinder misfire, random misfires, knock sensors, crankshaft and camshaft position sensors, ignition coil primary/secondary circuits, and ignition timing.

If a scanner detects that a PCM has set a P0300 "random/multiple cylinder misfire detected" DTC, or a P030X DTC where X represents a specific misfiring cylinder, the existence of the misfire should be verified before replacing expensive ignition parts. For example, on a coil-over-plug ignition system, if a P0352 Ignition Coil B Primary/Secondary Circuit code was set (this code is to indicate a bad ignition coil), there is a simple test that can be performed to confirm which ignition coil has set off the DTC. By swapping an ignition coil from the bad or potentially problematic cylinder that is setting the DTC via the

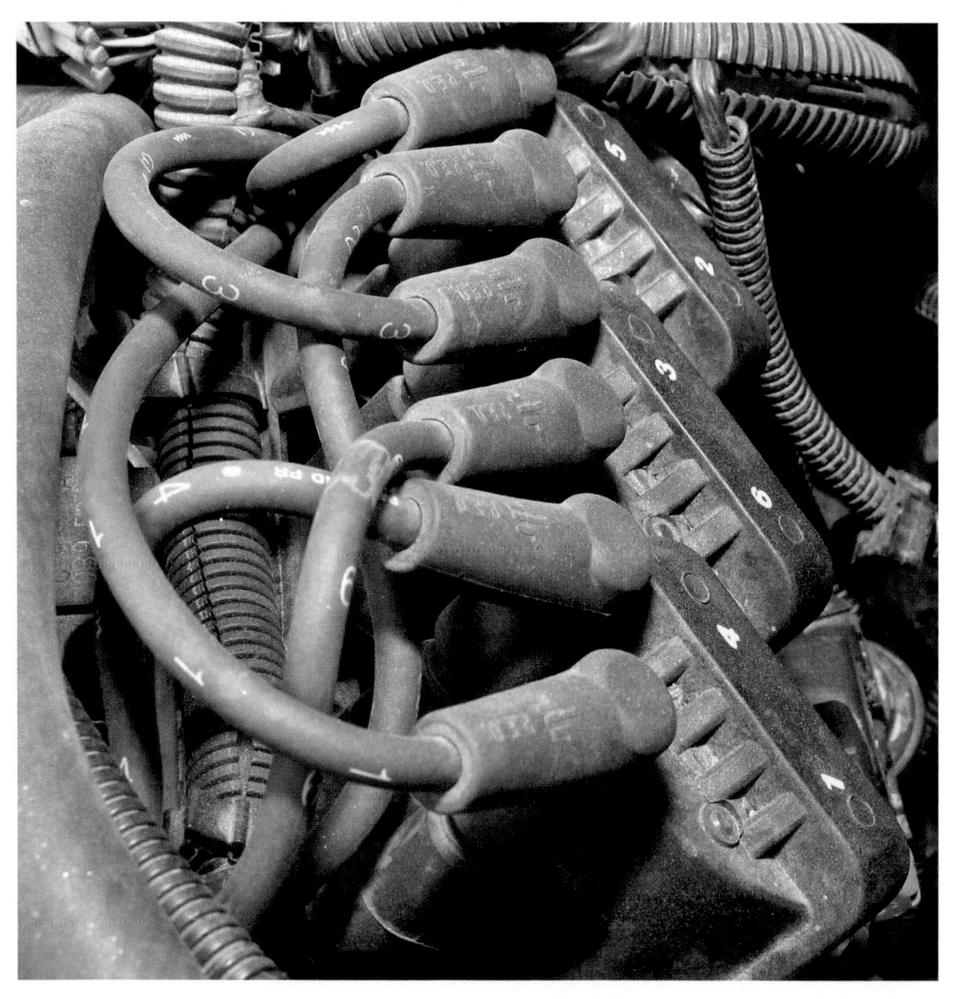

By 1990, many manufacturers had replaced ignition distributors with coil packs like the one pictured from a 1996 Pontiac. These systems were more reliable than earlier mechanical distributors, but they still used spark plug wires to transmit an ignition coil's high voltage energy to the spark plugs. Many of these systems have now been replaced by coil-over-plug systems that connect the ignition coil directly to the spark plug, a more advantageous design since there is no more spark plug wires that can wear out.

PCM, with an ignition coil from a good or perfectly operable cylinder, the bad coil can be indirectly tested. If a new DTC is set again for that same bad cylinder, despite the fact its original coil is now no longer installed on it, it can safely be assumed that the coil swapped out was good, and the wiring between the PCM and ignition coil or the PCM itself may be bad.

On all OBD-II vehicles equipped with coil-pack types of ignition systems, spark plug wires are still used and still suffer from the same problems experienced by older vehicles with plug wires. Since voltage traveling through spark plug wires is so high, it tries to make its way to battery ground via a path of least resistance, instead of traveling to the intended spark plug. This causes misfires and can set OBD-II-related trouble codes. Spark plug wires, terminals, and spark plug boots can all eventually wear out with age or otherwise become damaged; both conditions can cause arcing or voltage leaks.

When working on older OBD-I vehicles equipped with distributor caps and rotors, check the inside of the distributor cap for carbon tracking (which appears as shiny black lines). Additionally, green- or white-colored corrosion can also form on rotor tips and distributor cap terminals. A distributor cap and rotor can be checked using an ohmmeter, but this method of testing will only confirm that connections are not open, not how well they will work under actual operating conditions. Consequently, while these connections may check out all right with a meter, they can subsequently

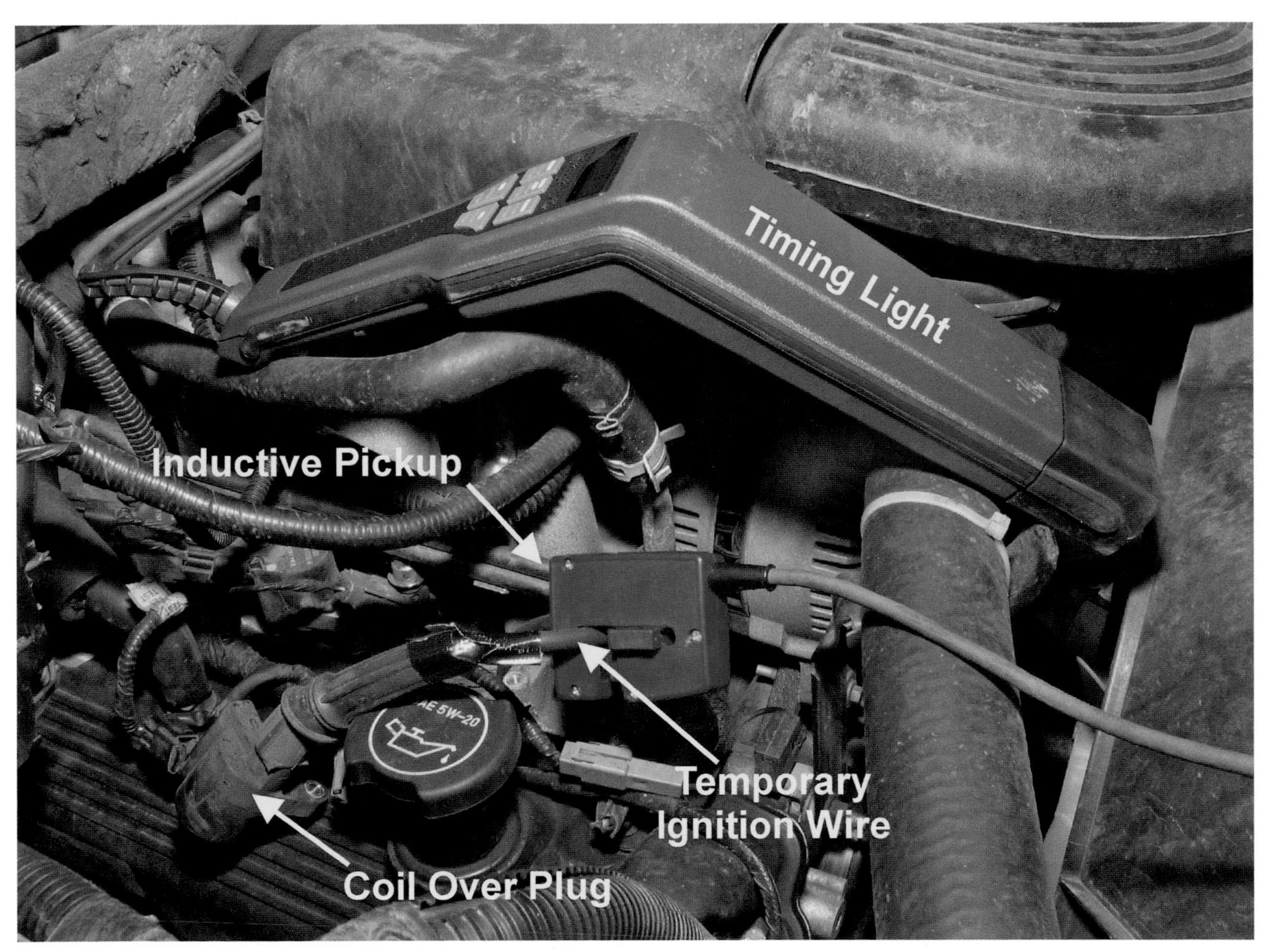

Any type of inductive timing light can be used as a poor-man's ignition scope. A coil-over-plug for one cylinder has been removed from this Ford truck, and a temporary ignition wire installed instead between the coil and spark plug. The timing light's inductive clamp is then placed over the temporary wire. With the engine running, the flashing timing light is observed. If a misfire occurs, the human eye and brain are fast enough to visually register it. This little trick also works to discover injector misfires as well. To find an injector misfire, the inductive probe is clamped over one of the two wires going to a fuel injector. If the PCM skips a beat and fails to fire the fuel injector, a momentary pause in the flashing timing light will occur, letting a technician know that injector has misfired.

After spraying secondary ignition wires with water, a grounded test light can be used to find bad wires or spark plug boots. When the sharp tip of a test light is placed close to a leaky wire, the test light will attract high-voltage spark energy that culminates in a misfire. This test also works well in the dark, as leaking ignition wires that are shorting high voltage energy to ground can be readily seen.

break down under load, causing ignition misfires. Spark plug wires should be visually inspected for cracks, brittle insulation, or loose connections at the distributor cap or spark plugs. Spark plug wires can also be checked using an ohmmeter, but resistance values vary widely between manufacturers, so be sure to consult an appropriate service manual for specific ohms per inch/foot values. Keep in mind that using an ohmmeter for resistance checking has limited value, because it can only detect if a wire is not broken, not how well it will perform with high voltage flowing through it. For example, when an engine's throttle is wide open, engine compression increases to maximum, which in turn places added electrical resistance on the secondary circuit, which likewise can cause a spark plug wire that previously checked out as good to turn bad and leak voltage to ground, and in the process, completely bypass the intended spark plug causing a misfire.

Using an inductive timing light to detect high-voltage ignition misfires is one of several low-cost methods that work on both coil pack and coil-over-plug ignition systems. Following is a brief explanation of how such a test works. When the pick-up clamp of a timing light is connected to a spark plug wire, the timing light will flash every time a spark is sent from the ignition coil to the spark plug. The human eye in conjunction with a timing light acts as a makeshift "poor man's" ignition scope that allows a technician to visually detect misfires. To check for ignition misfires, connect a timing light to the coil wire between the coil and distributor cap on vehicles equipped with distributors, or to the spark plug wire on ignition systems fitted with coil packs. On coil-over-plug systems, the coil must be removed and a temporary plug wire installed between the coil and plug in order to provide a place to connect a timing light for this test. Start the engine, and point the timing light at your face. The flashes from the timing light should be steady. If a misfire is bad enough, the flashes will skip or pause, indicating an ignition misfire. Now snap the throttle open momentarily, close it, and then open again to see if the light skips a few flashes. Sudden throttle opening should increase the resistance of the ignition circuit, which, in turn, should cause the coil to yield higher spark plug firing voltages. The increased voltage may cause a spark plug wire to go bad, or cause a coil or other ignition component to cause a misfire. If no misfire is detected with a timing light, but the engine is still obviously misfiring, move the inductive clamp of the timing light to each individual ignition wire (or temporary coil-over-plug wire) and repeat the test. Doing so will help isolate the component(s) causing the ignition to misfire.

Another method for finding bad ignition wires requires a 12-volt test light and a spray bottle filled with water. Use the spray bottle to really soak the ignition wires and/or components with water, since water helps magnify any high-voltage leaks from the ignition system. Connect the alligator clip of a 12-volt test light to battery ground and start the engine. Move the tip of the test light along all the ignition wires, around coil packs all the while listening to the engine. When the sharp tip of the test light is placed close to any wire or component that is leaking voltage, the high-voltage will jump to the tip of the test light, causing the engine to misfire. A variation on this test requires it to be duplicated in the dark, preferably at night. (CAUTION! Remember, having garage doors closed while an engine is running is not a good idea because of the risk of lethal carbon monoxide poisoning, which is undetectable by human senses.) In such a test, leaking high-energy voltage can often be seen and sometimes heard (listen for a spark arcing to ground) whenever the misfire occurs.

Fuel Injectors

Fuel injectors and their circuits are fairly reliable on modern vehicles, but on those infrequent occasions when they do misbehave, OBD-II systems have as many as 100 DTCs that can be set to address injector circuit malfunctions. Much like ignition system components, fuel injectors are expensive to replace; consequently, just because a DTC has been set doesn't mean you should rush right out and spend money on replacement parts, or possibly even costly repairs. Just as individual ignition circuits or coils can cause a PCM to set ignition-related DTCs, a shorted injector can do the same thing. To verify if an injector, PCM, or wiring harness is the cause of an injector code problem, try swapping injectors and then check to see if the same DTC follows with the injector to a different cylinder, or if it stays on the same cylinder where the DTC was originally set. For example, if P0204 DTC (Injector Circuit/Open—Cylinder 4) is set, swap the injector from a different cylinder and install it into the No. 4 cylinder, and install the original injector from the

Fuel injectors are simply electrical solenoids controlled by a vehicle's computer. Inside each fuel injector is a coil of wire and movable electromagnetic valve. When energized by an onboard computer, the coil of wire is transformed into an electromagnet, which causes the valve to open, thereby allowing fuel to be injected into the engine's intake manifold. A PCM controls the amount of fuel injected into the engine by varying the length of time the injectors are turned on, referred to as injector pulse width.

No. 4 cylinder into the other cylinder from which the injector was removed and clear all DTCs. Then, after performing an appropriate drive cycle, check to see if the OBD-II fuel monitor has completed its scan for any possible DTCs again. If the same DTC has again been set, but this time shows up on the cylinder where the original injector from cylinder No. 4 was reinstalled, then it is obvious that the original injector from cylinder No. 4 has caused the PCM to set the DTC, and not the wiring between the injector and PCM, or the PCM itself.

Another test that can be performed on fuel injectors measures its resistance using an ohmmeter. However, bear in mind these tests are not particularly sophisticated, in that electrical resistance readings will only confirm whether a specific injector's internal coil is not shorted to ground or open. Since resistance values vary widely between auto manufacturers, consulting an appropriate service manual for specific values for the particular vehicle being worked on is always a good idea. In fact, the only way to tell with certainty if an injector is actually working electrically is to use a lab oscilloscope to monitor the injector's pulse from a PCM. However, there are a number of simpler tests that will work and yield results most of the time which can be used to indirectly confirm whether an injector is operating properly, or at all, or not. Following are brief descriptions of four simple tests that can be used to confirm the operating condition of a fuel injector. The first three tests verify that a PCM is, in fact, sending an injector pulse signal to the injector, while the last test checks for and confirms mechanical injector operation.

Test 1. Unplug the injector and connect a 12-volt test light between the two wires at the injector connector

To check for the presence of an injector pulse from a vehicle's PCM, a test light is used in place of a fuel injector. The pointy end of the test light is touched to one of the wires in the injector harness, and the alligator clip end is connected to a straight pin, which is then inserted into the other injector wire at the harness. As soon as the engine is cranked over, the test light will flash or pulse if an injector signal is present at the connector. This simple test confirms whether a PCM is sending out an injector pulse to individual fuel injectors.

By removing the electrical connector from an injector and connecting a noid light to the injector's harness instead, an injector pulse from a PCM can be viewed once the engine is cranked over. Noid lights come with a variety of adapters to fit many popular injector wiring harnesses. They are a better choice than a test light to test for the presence of a PCM-generated injector pulse. Noid lights are available at many auto parts stores and online.

harness. Crank or start up the engine while watching the test light. If the test light flashes, the PCM is sending a pulse to that injector. (While a test light will work most of the time for performing this test, it's important to know this test will not work on all vehicles because some use a dropping resistor in the injector circuit that limits current going into the injector to keep it from overheating. See Test 2.)

Test 2. Instead of using a test light, use a noid light that is specific to the particular EFI system being tested. A noid light has a low enough resistance to flash during the test, even when a dropping resistor is used. Noid lights are available from some auto parts stores and on the Internet, and cost about $20.

Test 3. This test uses an inductive ignition timing light to verify injector pulse instead of a test light or noid light. However, the method performed is the same as the one used to check for an ignition misfire. Clamp the timing light's probe around one of the wires going to the injector. Start or crank the engine and watch the timing light to see if it flashes—a flashing light provides confirmation that the PCM is sending a pulse to that injector.

Test 4. This final test provides a low-tech method for confirming whether a fuel injector is receiving injector pulses from a PCM. To do this test, simply take a long screwdriver and touch the end of it to a fuel injector and stick the handle end in your ear. (No kidding!) If the injector is working, you should be able to hear a steady clicking from the injector as sound waves from the injector opening and closing are transmitted through the screwdriver. A wooden dowel, mechanic's stethoscope, or even a simple piece of vacuum hose will also work for this test.

No-Start Conditions

In those instances when an engine cranks over, but just won't start, there are some basic things any technician, regardless of skill level, can check or do before calling a tow truck. The first thing is obvious: use a code reader or scan tool to check if any DTCs have been set by the PCM. Diagnostic trouble codes are a good place to start any investigation into potential causes of a specific problem, since they may provide valuable clues as to what the problem is, or where the malfunction is most likely located, or at a minimum, in what system the problem is most likely occurring. For instance, if any codes in the range of P0350 to P0362 are set, there is most likely a problem with the vehicle's ignition system that is causing a no-spark condition. However, P02XX fuel and air metering codes are also

A mechanic's stethoscope (available at most auto parts stores) is used to listen to the operation of a fuel injector. If working properly, the injector should make a steady clicking sound and speed up in direct correlation with engine speed. If the PCM is not sending an injector pulse to the injector, or the pulse is intermittent, the stethoscope will allow the technician to hear this problem.

A short section of 3/8-inch hose can also serve as a satisfactory method for listening to a clicking fuel injector. In the photo, the author is actually speaking into the hose to test if the sound waves from his voice (like sound waves from a fuel injector) will travel through the hose into his ear—true science in action!

a possibility, so these should be checked. Moreover, remember, just because a code has been set by a PCM doesn't mean it's time to start replacing parts, at least not until it has been verified by other independent methods that a no-start condition really does exist. Keep in mind, ignition system DTCs only provide a clue to the source of the problem, not necessarily the answer.

As most technicians know, slow engine cranking speed will prevent an engine from starting. Engine cranking speed should typically be around 200 rpm, but engine speeds are actually difficult to measure at this level without special equipment. Fortunately, with some experience, a person's ear will suffice as a basic test for determining engine-cranking speed. If an engine is obviously cranking slowly, there will be insufficient compression generated inside the cylinders to promote combustion, regardless of whether an ignition spark and fuel are present. Slow engine cranking can be caused by a bad or discharged battery combined with cold weather, loose or poor battery cable connections, a bad starter motor, or other mechanical engine problems.

Missing in Action—Spark or Fuel?

If an engine is cranking at normal speeds, but the engine still won't start, then the big question that needs to be answered is: Is the engine missing spark or fuel or both? It is easy to find out why an engine is failing to start, and the method used to find out does not require having to take anything apart. Instead of wasting time removing ignition coils and/or spark plug wires to check for an ignition spark, try this simple test that completely bypasses the entire fuel injection system. Take a hose from a propane bottle and

Using propane is a relatively safe way to determine if an engine has spark, but no fuel. This simple test bypasses the entire electronic fuel injection system. To perform the test, simply insert the hose from the propane bottle into the intake duct at the throttle body, then turn on the bottle's valve and crank the engine. If the engine starts and runs (even for only a few seconds), then the presence of an ignition spark is confirmed. The valve on the propane bottle shown in the photo is actually designed for automotive use, so it is equipped with a plunger at the top that instantly opens the valve. Propane flow is controlled via the round valve located at the top of the propane bottle (other valves will also work, but not as well).

either snake it into the air cleaner or directly place it onto the throttle body. Then, open the propane bottle's valve slightly while a friend or assistant cranks the engine over. If the engine starts and runs, even for a few seconds, the ignition system is producing a spark.

In such an event, be sure to check for the presence of 12 volts at all fuses (especially the fuel pump fuse), as less than full power could cause no-fuel problems. Furthermore, if an engine doesn't start even after a shot of propane, then a check for ignition problems is in order. Also be sure to check ignition/coil power-related fuses in addition to fuel system-related fuses. Although a nonoperational fuel pump won't necessarily cause a PCM to set a DTC, a good place to start an investigation is to check the P0230 to P0233 range of DTCs (all are related to fuel pump primary circuit operation) as the cause of a no-fuel condition. Most vehicles' onboard computers are programmed to allow a fuel pump to run for a few seconds after the ignition key is first turned on to prime the fuel system. Have a friend or assistant turn the ignition key to ON while you place an ear to the open fuel tank filler hole—you should be able to hear the fuel pump running. However, bear in mind that not all vehicles may allow a fuel pump to run (check an appropriate vehicle service manual to verify this fact), so don't automatically assume there is something wrong with the fuel pump if you don't hear it running when the key is first switched to the ON position.

Once it is verified that both spark and fuel are present in a combustion chamber, but an engine still won't start, use a vacuum gauge to check if cranking vacuum exists, since cranking vacuum is an indication that the exhaust is not plugged up or that camshaft timing is correct—both of these can cause a no-start condition (see Figure 6-10, p. 128). If valve train timing is suspected (one indication of no-cranking vacuum), camshaft timing can be verified by inspecting the alignment marks on the crankshaft and camshaft. A service manual will have to be consulted to locate the timing marks and interpret correct engine valve timing alignment. Finally, the absence of cranking vacuum can also be caused by a plugged exhaust system, which can occur if a vehicle was driven over a curb and it smashed the exhaust pipe, or a plugged catalytic converter. To test for a plugged converter, disconnect the exhaust pipe located between the catalytic converter and the engine by simply loosening any exhaust connection sufficiently enough to allow exhaust gases to bypass the catalytic converter. If the engine now starts and runs, the source of the problem has been discovered.

Another cause for a nonstarting engine could be any related vehicle alarm codes. Fuel-enabled or alarm-related DTCs are not generic OBD-II DTCs, but are instead specific manufacturer's enhanced trouble codes. For example, any of the antitheft-related DTCs could actually prevent a PCM from starting the engine. A PCM will do this simply by shutting off components such as a starter motor, fuel injectors, or ignition coils. A specific example of this can be seen on certain Chrysler vehicles, where an enhanced manufacturer-specific OBD-II trouble code (Fuel Enabled = NO) exists. On these vehicles, a PCM sometimes reacts in a manner consistent with having an alarm activated, even if no alarm exists on the vehicle in question. In such instances, the problem may actually be with the PCM itself, in that either a wrong PCM has been installed, or a specific PCM needs updated programming.

Now that we've covered basic engine diagnostic procedures we'll move on to how to use a scan tool to diagnose basic OBD-II-related problems and trouble codes, putting what you know to practical use.

CHAPTER 7
SCANNER OPERATION

This chapter will discuss how to effectively use a scan tool to diagnose OBD-II-related problems. For those of you who skipped Chapter 6, Automotive Detective Work, you will probably have to read it anyway because the engine in your car or truck is still, after all, just a mechanical air pump that has to seal and operate correctly. If there is a mechanical problem with the engine, the PCM will not have the ability to diagnose basic engine-related problems such as low compression, vacuum leaks, burned or bent valves, or blown head gaskets. In addition to engine mechanical problems, any issues with a weak or bad battery, or charging system-related problems will cause unreliable diagnostic results from the PCM. These results may show up as multiple DTCs that seem to have no direct relation to each other. These types of DTCs may show up as sensor signal outputs that are way out of range and/or PCM output failures. Chapter 6 shows you how to separate OBD-II computer-related issues from engine mechanical and charging system/battery problems. So before you tackle the service-engine-soon light, and related DTCs, make sure the problem is not something low-tech or simple that has been overlooked.

Start the diagnosis process by reading any DTCs that the PCM may have stored. If the vehicle is having drivability problems, check for codes even if the MIL is not on, as there may be pending DTCs present. If the MIL is on, check to see if it also flashes when the vehicle is being driven. The behavior of the MIL can provide clues as to what's wrong. Also note when any drivability symptoms occur. Is the engine hot or cold? Does the vehicle have to be driven for a short or long period of time before the problem shows up? How long have the symptoms been present—did they just happen right after someone else replaced a component or worked on the vehicle?

If there are any trouble codes stored in the PCM, check for freeze-frame data. If your scan tool has the capability, print out freeze-frame data and any codes scanned. If the scan tool or code reader cannot print these test results, write down the freeze frame. Don't just read and then erase DTCs that are present in the PCM's memory. On most vehicles, when the codes are cleared so is all the freeze-frame data, and if you inadvertently lose this valuable information it may take considerably longer to correctly diagnose what's wrong. Make a visual inspection of the engine compartment and look for anything that is obviously disconnected or damaged, including vacuum hoses and sensors. Especially pay attention to areas indicated by any DTCs that have been set by the PCM. For example, if one or more DTCs present are related to the EGR system operation and you find a loose, cracked, or disconnected EGR vacuum hose, you may have solved your problem. The PCM and all of its inspection and maintenance monitors can lead you in the general direction of where to look for a problem, but often does not provide specific enough information to get to the root of engine performance or emission-related issues. Remember that with all its computing power and sophistication, the PCM is not as smart as you, so don't let it take you in the wrong diagnostic direction and in the process waste your time and money.

Trouble Codes, OBD-I vs. OBD-II

When a trouble code is present, OBD-II offers much more specific information related to the faulty system or component than earlier OBD-I systems did. For example, trouble code 53 on an EEC-VI Ford from the mid-1980s was defined as throttle position sensor malfunction. Numerous problems and conditions could cause this code to be set including: disconnecting the sensor and then turning the ignition on, a problem with the wiring harness or connectors between the TPS and the electronic engine control unit (onboard computer), a bad computer, loose TPS sensor, misadjusted throttle plate, electrical interference from secondary ignition wires, and last but not least, a bad throttle position sensor. Many a technician wrongly assumed that Ford code 53 indicated that the TPS should be replaced and more often than not this did nothing to fix the underlying problem. Scanning trouble codes on an OBD-I system only provided very general clues, and not much more, as to where to start looking for drivability or emission-related problems.

OBD-II has made drivability and emissions diagnosis much easier by providing more specific, detailed information when a malfunction occurs. In the case of the throttle position sensor, there are five generic OBD-II trouble codes and seven manufacturer-specific (Ford in this case) DTCs, all for the throttle position sensor and its electrical circuit

	DTC	Code Definition
OBD-II Generic Codes	P0120	Throttle/Pedal Position Sensor/Switch "A" Circuit
	P0121	Throttle/Pedal Position Sensor/Switch "A" Circuit Range/Performance
	P0122	Throttle/Pedal Position Sensor/Switch "A" Circuit Low
	P0123	Throttle/Pedal Position Sensor/Switch "A" Circuit High
	P0124	Throttle/Pedal Position Sensor/Switch "A" Circuit Intermittent
Manufacture Specific OBD-II Codes	P1120	Throttle Position Sensor "A" Out of Range Low *While the engine is running, the PCM detected the TPS voltage was below its designated spec.*
	P1121	Throttle Positions Sensor Inconsistent with Mass Airflow Sensor *While the engine is running, the PCM detected that the TPS and the MAF sensor are not consistent with their designated values. Check the Real Time data, TPS = < 4.8% @ a Load of 55% or TPS = > 49.05% @ a Load of 30%.*
	P1122	Throttle Position Sensor "A" in Range but Lower Than Expected *While the engine is running, the PCM detected the TPS to be below its designated spec.*
	P1123	Throttle Position Sensor "A" in Range but Higher Than Expected *While the engine is running, the PCM detected the TPS to be above its designated spec.*
	P1124	Throttle Position Sensor "A" Out of Self-Test Range *While the engine is running, the PCM detected the TPS to be out of its designated spec of less than 13.27% with KOEO, or more than 49% with KOER.*
	P1125	Throttle Position Sensor "A" Intermittent *While the engine is running, the PCM detected the TPS to be out of its designated spec. Check the Real Time data, and do a wiggle test to look for intermittent changes in the output.*
	P1126	Throttle Position Sensor Circuit *While the engine is running, the PCM detected the TPS to be out of its designated spec. Check the Real Time data, and do a wiggle test to look for intermittent changes in the output.*

Fig. 7-1—OBD-I systems usually only provide a single (or maybe two) trouble code(s) if anything is wrong with its system computer sensors. OBD-II has much greater diagnostic capabilities. In this example, there are five generic DTCs and seven manufacturer-specific trouble codes—all for a throttle position sensor. OBD-II provides this detail level of diagnostic information that will hopefully narrow down what is actually wrong with the TPS sensor, or its related electrical circuit.

While some of the TPS messages in this picture are from older OBD-I systems, they still provide the same information as they do on newer OBD-II-compliant vehicles. Fortunately when something goes wrong with a TPS on an OBD-II-system, the numerous TPS-related trouble codes can help pinpoint where a potential problem lies.

(see Figure 7-1). Using only the generic OBD-II TPS-related DTCs the following information about the TPS and its electrical circuit is available: TPS output voltage is too high or low, TPS range of output voltage does not match other sensor outputs, and even intermittent circuit problems can be pinpointed simply by reading these OBD-II generic TPS-related trouble codes.

The manufacturer-specific P1 codes provide even more detailed TPS circuit information. Code P1121 is defined on a scan tool as "Throttle position sensor inconsistent with mass airflow sensor." Looking in a service manual, a more detailed code P1121 definition states "While the engine is running, the PCM detected that the TPS and the MAF sensor are not consistent with their designated values. Check the Real Time data, TPS = < 4.82% @ a load of 55% or TPS = > 49.05% @ a load of 30%." This means that the PCM has received a TPS signal that is less than 4.82 percent of the throttle opening with a calculated load value (from the MAF sensor) of 55 percent. Because the engine load value is high at 55 percent (the engine is working at 55 percent of its total capacity), the TPS should have been sending an output voltage signal to the PCM indicating that the throttle was opened considerably more than 4 or 5 percent. The other half of this code definition (TPS = > 49.05% @ a load of 30%) is the opposite of the first. The TPS output voltage signal to the PCM is indicating that the throttle is open 49.05 percent but the calculated engine load value is only at 30 percent—far too low for the throttle being open half way, or 49 percent. The TPS-related DTCs in this example are typical of the detail of diagnostic information that OBD-II systems provide via a scan tool or code reader. Always keep in mind that while the P1121 code could be caused by a bad TPS sensor, it could also be caused by a loose sensor or misadjusted throttle plate. The P1121 is a specific clue, but only a clue and not necessarily the ultimate answer to a drivability or emission-related problem.

Live Data Stream, or PIDs

By checking the parameter identification (PID) information on a scan tool and comparing the PCM data values with a service manual, it can readily be seen when a sensor is out of range or inconsistent with other sensor inputs. For example, if the engine had just been started and running for only a minute, the coolant temperature sensor (CTS) reading should be higher than the intake air temperature (IAT). Because the engine is running, the CTS should increase to normal operating temperature (over 200 degrees Fahrenheit) within a few minutes at idle, or sooner by driving a short distance. If the CTS value was the same, or only slightly higher than the IAT, or if the coolant temperature was lower than the IAT, a problem is indicated with the coolant sensor. In addition, be careful in using generic, seat-of-the-pants sensor ranges and

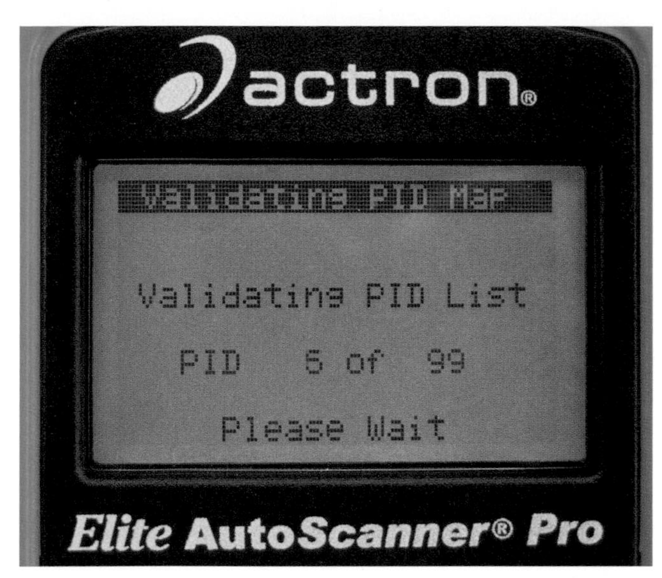

When "View Data List" is selected from the main menu of this Elite AutoScanner Pro, the scan tool accesses the PCM and validates the PID list. PIDs are nothing more than the data stream from the PCM and its various sensors. The vehicle that the scan tool was connected to had 99 PIDs available. *Courtesy Actron*

The live data stream from the vehicle that this Actron scanner is connected to is being displayed. By using the down arrow key on the scan tool, more PIDs, or sensor data, will be displayed. Comparing PID values to those found in a service manual is a good way to verify that sensor outputs are where they belong and that nothing is out in left field and causing a problem. For example, the engine in this vehicle has only been running for a few minutes. The coolant temperature sensor reads 20 degrees higher than the intake air temperature sensor, indicating that it's functioning properly. *Courtesy Actron*

Jeep Grand Cherokee
Technical Service Bulletin #18-40-98

18-40-98 -- HIGH IDLE RPM DURING DECELERATION
Date: 12/18/98

Model year(s): 1999

Description: THIS INFORMATION APPLIES TO VEHICLES EQUIPPED WITH 4.01L ENGINES BUILT PRIOR TO DEC. 18, 1998. Some vehicles may exhibit high idle RPM during deceleration and/ or parking lot maneuvers when the engine coolant temperature is greater than 103 c (217 f). The increased idle rpm will return to normal when the vehicle comes to a stop. Coolant temperatures greater than 103° C (217° F) are more noticeable in high ambient temperatures during stop and go traffic, or after towing.

Details: Using the Mopar Diagnostic System (MDS/MDS2) or the Diagnostic Scan Tool (DRB III®) with the appropriate Diagnostic Procedures Manual, verify all engine/transmission systems are functioning as designed. If Diagnostic Trouble Codes (DTC's) are present, record them on the repair order and repair as necessary before proceeding further with this bulletin. Make sure the cooling system is working properly and verify that the engine temperature sensor values are valid. If no other problems are found, reprogram the PCM following the proper Repair Procedure as outlined on page two of this TSB.

Fig. 7-2—Pictured is an example of a typical technical service bulletin. This one is for a Jeep Grand Cherokee model with a 4.0-liter engine, built prior to December 18, 1998. The symptoms are a high engine idle speed when the engine temperature is above 217 degrees Fahrenheit. In addition, the TSB states that idle speed will return to normal when the vehicle comes to a stop. The Details paragraph advises the use of a Mopar Diagnostic System or factory diagnostic scan tool (DRBIII) to verify that all systems are functioning as designed and to read any DTCs that are stored in the PCM. The repair procedure to fix this problem is to reprogram or flash-update the PCM. What the TSB doesn't state is that the only way to accomplish this is by use of a factory scan tool or diagnostic system. This TSB also illustrates that the cause of some drivability problems are not related to components that are broken or malfunctioning. In this case a software update is the only way to fix the problem. A technician unaware of the existence of this type of TSB could spend needless hours trying to find something wrong with the vehicle.

values, as these may vary widely between manufacturers and even between model years and engine families within the same manufacturer.

Where to Find Information

In addition to scanning for DTCs and reading a vehicle's data stream, it's always worth checking for technical service bulletins (TSBs) from the manufacturer of the vehicle you are working on. This is especially true if the MIL is on and there are no drivability issues. Often trouble codes or other weird drivability problems are just a matter of updating the PCM's software with the latest fix from the factory. Reprogramming the PCM is typically done on new vehicles because sometimes automotive engineers who design these systems don't always get things right the first time, and after a model has a few thousand miles on the odometer design flaws may show up. TSBs and other service information

1999Audi A4 Avant: DTCs P0301/P1300

Display groups: 15 and 16 Display Fields: 1-4	Possible cause	Corrective action
larger than 5	• Ignition coil faulty • Spark plug connector faulty • Spark plug faulty • Power output stage for ignition coil faulty	– Check ignition coil – Check power output stage for ignition coil
	• Fuel injector faulty • Not enough fuel	– Check fuel injectors actuation – Check fuel in tank

Today, even the most experienced technicians rely on subscription-based internet information clearinghouses, such as Alldata.com. Thousands of mechanics from all kinds of shops share their knowledge on these websites, making it much faster and easier to diagnose problems causing DTCs. In this case, the second suggested corrective action proved to lead to the problem's solution. Alldata.com isn't just for pros; they have a DIY subscription service, too.

A woman was using her 1999 Audi A4 Quattro Avant to run some errands for work. She had been driving the car for a number of relatively short periods in succession when the car suddenly started vibrating. She checked the dashboard display and noticed the check engine light blinking. The engine, a turbocharged 1.8L (commonly known as a 1.8T), was running roughly. She made it to work, where she left the car parked for the rest of the day. When the car was restarted and driven home, which only took a few minutes, it ran fine. The next morning, the car ran fine upon startup, then resumed running roughly after warming up a bit.

Her husband, armed with an Actron PocketScan code reader, retrieved two diagnostic trouble codes, or DTCs. The two codes were P0301 and P1300. P0301 is a generic code indicating a misfire in cylinder number one. P1300 is an Audi-specific code indicating an intermittent misfire for the entire ignition system.

The car was taken to Good Carma, an independent repair shop in Minneapolis that specializes in Volkswagens and Audis. One of their technicians, Brian Archer, took a look at the car's diagnostic system using the Volkswagen Auto Group (VAG) scan tool. This is a Windows-based emulator and diagnostic tool that acts as a scan tool but has additional capabilities specific to Volkswagen and Audi vehicles. The VAG scanner found the same two DTCs as the PocketScan code reader—a multiple random ignition misfire and a specific ignition misfire in cylinder number one.

Brian wanted to find out if the misfire in cylinder number one was related to its ignition coil pack (this engine uses individual coil packs for each cylinder). He swapped the number one cylinder's coil pack with that of cylinder number four. After starting and running the engine for a few minutes, he read the DTCs and found the same codes indicating random multiple misfires and misfires in cylinder number one. This eliminated the coil pack as the problem, since the misfire did not move to cylinder number four.

The next step was to swap the spark plugs between cylinders one and four to determine if the misfire was caused by a faulty spark plug. As before, the misfire stayed with cylinder number one. With the cylinder number one spark plug and coil pack both eliminated as the cause of the misfire, only two possibilities were left—low compression in cylinder number one or a bad electrical signal from the Audi's powertrain control module (PCM) to the number one coil pack. A compression tester was used to verify that engine compression in cylinder number one was normal. That left only the PCM as the cause of the misfire—or did it?

From his experience working on Audis, Brian knew that this car's ignition system uses the PCM to generate an ignition signal based on various sensor parameters. The PCM's ignition signal is sent to what Audi refers to as a "power output stage," which is basically an ignition module. The power output stage provides a ground for each of the ignition coil packs, in the correct firing sequence, completing the circuit and firing their respective spark plugs. If the PCM's signal is making it to the power output stage but the power output stage isn't firing the ignition coil, or is firing the wrong coil, then the power output stage needs to be replaced.

Brian used a logic probe, sometimes called a red-and-green LED test light, to check for the presence of an output signal from the PCM through the power output stage to the number one ignition coil. With the engine running and the LED test light connected to the coil pack's input wires, the light did not flash, indicating no signal. A new power output stage was installed, the DTCs were cleared from the PCM, and the car was test driven. The engine operated smoothly, no more ignition misfires, and the check engine light remained off.

This diagnostic story illustrates how a basic code reader can give a car owner an idea of what the problem is, but sometimes it takes an experienced professional technician to determine what component is actually at fault.

can be found in a variety of places. Sometimes you can find TSBs on the manufacturer's website or a discussion forum where owners of the same vehicle share technical information and tips and tricks for their favorite car or truck. Actron and AutoXray have a subscription website called CodeTrack where do-it-yourself technicians can access specific information on OBD-II codes and TSBs. The website address for CodeTrack is www.codetrack.spx.com. Another good source of information is the International Auto Technicians Network (IATN). With over 50,000 members as many as 2,000 technicians may log on to the website at one time—all sharing information and giving advice. The IATN website is located at www.iatn.net.

Finally, if you want to talk live to an actual technician about your specific drivability or OBD-II related problem, try over-the-phone SCM Hotline. This resource currently supports a large number of professional and home technicians and has quick access to technical data from the Hotline's in-house database. Vehicle coverage includes both foreign and domestic cars and light trucks (gasoline and diesel) from 1964 to present. The automotive systems that are covered include: engine performance, OBD-I/OBD-II diagnostics, climate control (HVAC), ABS braking systems, computer-controlled transmissions, supplemental restraint systems (airbags) and electronic body control systems. Information regarding OEM factory information, technical service bulletins, wiring diagrams, repair procedures, component locators, and other service-related information is available. The hotline can be reached at 800-847-9454 or on the Internet at www.AutoHotLineUsa.com.

Using OBD-II Drive Cycles and Diagnostic Monitors

As discussed in Chapter 2, OBD-II system monitors can be used to diagnose both emissions and drivability problems. There are two basic reasons for taking this approach. The first is that unlike OBD-I systems of the past, OBD-II is robust enough to do a pretty fair job in figuring out what is working, or not, with the emissions systems that it monitors. Because many emissions-related systems and controls are closely associated with engine performance (fuel injection and ignition systems are but two examples) these systems' ability to function properly will have a great effect on drivability and engine performance issues as well. OBD-II diagnostics have been designed, developed, and installed on all vehicles since 1996, so why not take advantage of its diagnostic capabilities. By allowing the OBD-II system to run all its I/M monitors a technician can have reasonable (but not absolute) confidence that everything is working as it should.

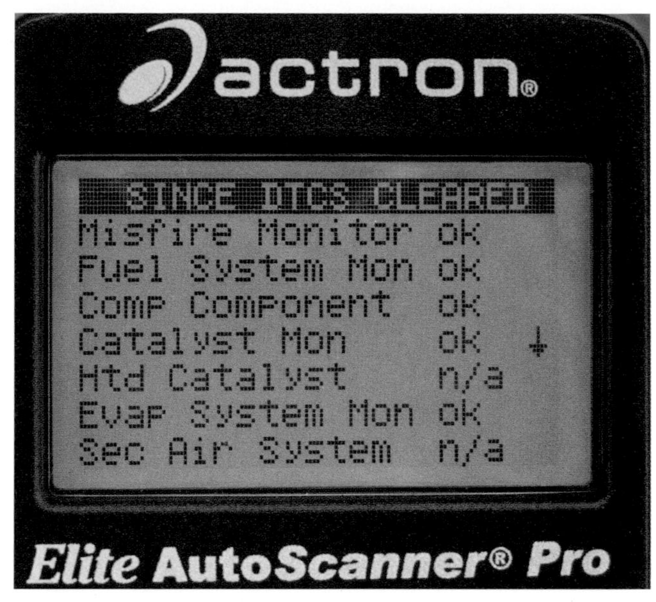

This Actron scanner is displaying the status of I/M monitors. This information verifies that the PCM has tested all of the emission systems on the vehicle. The status of the monitor displayed is since the DTCs were last erased, or cleared by the scan tool. Monitors are indicated as "OK" when they have completed, "Inc." when not completed and "N/A" when that system is not present on the vehicle. To get all the monitors to run and complete, a manufacturer's drive cycler will have to be performed. *Courtesy Actron*

The second reason to use the OBD-II system for diagnostic work is its ability to both turn off the MIL and to verify if any repairs that were made were successful. Clearing trouble codes and turning off the MIL can be accomplished using a scan tool or code reader. However, if a component was replaced because a DTC was set, one of the best ways to know that the problem is actually fixed is to see if the vehicle's PCM can turn off the MIL by itself. Using the PCM to turn off the MIL has another advantage. Turning off the MIL using a scan tool can be a problem if the vehicle is directly driven to a state emission testing facility for a smog test. Here's why: When DTCs are cleared using a scan tool, or code reader, all I/M monitors are set to "not ready" or "incomplete." Most state testing programs require that most, if not all I/M monitors be in a "ready" state to pass emissions testing. The vehicle will have to be driven enough random miles to cause the PCM to run all I/M monitors before the vehicle can pass emission testing. As an alternative to driving many miles, performing the manufacturer's specified drive cycle will speed up this process and set the I/M monitors to a completed, or ready, state sooner than merely driving the vehicle.

1998 Dodge Caravan: "U" Code

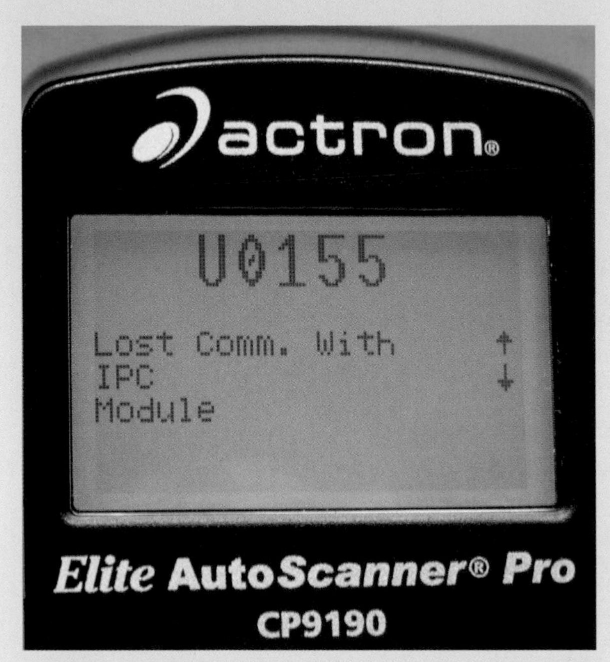

OBD-II vehicles that use the controller area network, or CAN, protocol have a set of diagnostic trouble codes with "U" as the first digit. The "U" trouble codes identify a network communications problem. In the case of this Dodge minivan, for example, U0155 is the code for a loss of communication between the instrument panel cluster and other controllers on the CAN network.
Courtesy Actron

Have the dash lights and instruments mysteriously shut off on your 1998 Dodge Caravan 3.8L minivan? Connect a scanner or code reader to access the OBD-II system and retrieve the diagnostic trouble codes, or DTCs. On some of these vehicles you may download a "U" code. These types of codes indicate that there may be a problem with the vehicle's controller area network, or CAN, system. "U" codes cover everything from vehicle communication performance to lost communication between CAN modules along the BUS (network) wire.

In the case of the 1998 Caravan, a U0155 is defined as "Loss of Communication with Instrument Panel Cluster (IPC)." This means the instrument cluster CAN module has lost communication with the vehicle's powertrain control module, or PCM. A smack on the center top of the dash may bring the instrument cluster temporarily back to life. If so, it also indicates that a connector is probably loose. (See Chapter 2, page 35 for more information on CAN.)

By following a specific manufacturer's drive cycle to run all the continuous and noncontinuous I/M monitors, the OBD-II system will do a fairly extensive test of fuel, ignition, and other engine management systems. Chapter 2 had an example of a generic OBD-II drive cycle, but the specific manufacturer's procedure is a better way to ensure that all the monitors will run. A scan tool or code reader connected to the OBD-II system can be used to see exactly when a particular monitor has been completed. A word of caution: many of these I/M monitors can be verified by reading data from a scan tool during the drive cycle. Needless to say, looking at the PIDs (live data stream) on the scan tool's display and watching the road at the same time is not a good idea or a very safe driving practice. Have a helper along for the ride when you perform drive cycle monitoring and testing. The following are drive cycles for General Motors, Ford, Daimler/Chrysler, and European vehicles. Some vehicles may require specific drive cycles instead of manufacturer-generic drive cycles for their monitors to run. A service manual should be consulted to determine the exact drive cycle to be used.

General Motors Drive Cycle

The General Motors drive cycle will test all the emission-related components and their systems. This drive cycle tests various groups of I/M readiness monitors instead of individual monitors. Drive cycles of specific General Motors vehicles may be different than the one presented here. A service manual should be consulted for drive cycle procedures for specific models. Here is a generic drive cycle for General Motors vehicles.

1. **Cold Start**—In order to be classified as a cold start, the engine coolant temperature must be below 122 degrees Fahrenheit and within 11 degrees Fahrenheit of the ambient air temperature at startup. Do not leave the key on prior to cold-starting the engine or the heated oxygen sensor monitor may not run successfully.
2. **Idle**—The engine must be running for 2 1/2 minutes with the air conditioner and rear defroster turned on. These systems apply an electrical load to the alternator. In fact, the more electrical loads you can apply to the charging system the better for completing the OBD-II monitors. This will test and run monitors for the O_2 heater, passive air, purge EVAP flow, ignition misfire, and if closed loop is achieved, fuel trim.
3. **Accelerate**—Turn off the air conditioner and all the electrical loads. Apply half throttle and accelerate the vehicle until 55 miles per hour is reached. During this time, the misfire, fuel trim, and purge flow monitors will be run.

General Motors drive cycles vary between its family of cars and light trucks. A service manual should be consulted for specific drive cycle procedures that apply to the OBD-II vehicle being tested.

4. **Hold Steady Speed**—Hold a steady driving speed of 55 miles per hour for three minutes. During this time the O_2 response, secondary air, EGR, canister purge, misfire, and fuel trim monitors will be performed.
5. **Decelerate**—Let off the accelerator pedal. Do not shift or touch the brake or clutch. It is important to let the vehicle coast along, gradually slowing down to 20 miles per hour. During this time, EGR, purge, and fuel trim monitors will be run.
6. **Accelerate**—Accelerate the vehicle at 3/4 throttle until 55 to 60 miles per hour is reached. This step will run the same monitors as in Step 3.
7. **Hold Steady Speed**—Hold a steady speed of 55 miles per hour for five minutes. During this time the same monitors from Step 4 will be performed, only this time the catalyst monitor diagnostics will be performed as well. If the catalyst is marginal or the battery has been disconnected, it may take five complete driving cycles to determine the state of the catalyst and to set its monitor to a complete, or ready, status.
8. **Decelerate**—This step will perform the same monitors as in Step 5. Don't press the clutch or brakes or shift gears during this last part of the General Motors drive cycle.

Ford Drive Cycle

Ford Motor Company's OBD-II drive cycle differs from General Motors in that individual inspection and maintenance monitors can be run separately from the overall drive cycle. On many Ford vehicles, manufacture code P1000 (EVAP system) will be set if all the I/M readiness monitors are not complete. Clearing the P1000 can be a bit tricky because the evaporative emission monitor test has to be completed before the PCM allows the P1000 to be cleared. The problem is that the EVAP monitor will only run during the first 30 minutes of engine operation from a cold start condition. There is an EVAP monitor bypass procedure but it only amounts to letting the vehicle sit for eight hours before trying to run the monitor again. The following four procedures will prepare the vehicle for monitor testing and clearing the Ford P1000 DTC.

Most OBD-II monitors will complete more readily using a steady-foot driving style during cruise or acceleration modes. Operating the throttle in a smooth fashion will minimize the time required for monitor completion. This will also minimize fuel slosh in the gas tank, which allows the EVAP monitor test to run successfully.

The ambient air temperature should be between 40 and 100 degrees Fahrenheit and altitude should be below

Ford Motor Company's OBD-II drive cycle when run correctly will clear any diagnostic trouble codes and turn off the MIL. The fuel tank should be three-quarters full for the EVAP monitor to run successfully, and the ambient air temperature should be between 40 and 100 degrees Fahrenheit.

Gas Cap Loose: DTC P0442

Diagnostic trouble code P0442 is a common DTC and is triggered when the PCM detects a small leak in the EVAP system. This often occurs when the owner neglects to tighten the gas cap sufficiently after a fill up. The gas cap pictured states "Tighten to 3 clicks," which is necessary to properly seal the EVAP system. *Courtesy SCM Hotline*

How many times have you pulled into a gas station and found a gas cap on top of a pump? On many newer vehicles, the gas cap is attached to the filler spout, but gas caps on older vehicles, and some newer ones, are not attached. With pre-OBD-II vehicles, leaving your gas cap behind (or not tightening it all the way) only resulted in some fuel leaking onto the road. Drivers didn't know there was a problem until they stopped and saw fuel dribbling out the filler door. This changed with the introduction of OBD-II in 1996.

Among other things, OBD-II systems monitor the vehicle's evaporative emissions system, or EVAP, to detect if any leaks are present. Since federal emissions standards are at odds with raw hydrocarbons (gasoline) escaping into the atmosphere, the integrity of the EVAP system is constantly monitored by the powertrain control module, or PCM. If your "check engine" light comes on as you drive away from a recent fill up, you might not have tightened your gas cap sufficiently—or you might even have left it on top of the pump. Also, if dirt or debris gets on the gas cap's O-ring, it needs to be wiped off before the cap is used to seal the gas tank or the P0442 DTC may be set.

P0442 (Evaporative Emission System Leak Detected; Small Leak) and P0455 (Evaporative Emission System Leak Detected; Gross Leak/No Flow) are the two most common DTCs set by the PCM for gas cap leaks and loose or missing caps. The P0442 small leak is more common, and it usually indicates that the gas cap is simply loose and needs to be removed and tightened again. Fixing the problem (tightening the gas cap) is easy; however, turning off the "check engine" light requires a bit more effort. If you don't have a code reader or scanner, you can drive the vehicle for the required number of duty cycles (this could take a week or more, depending on how often you drive it) with the check-engine light on until the OBD-II system turns it off, or you can take it to a dealer.

General Motors recognized this to be a problem and, starting in 2002, added a new warning light to the driver's information display on the Cadillac Seville. When the gas cap is left loose, a new "Check Gas Cap" warning indicator lights up without setting a trouble code. Pull over and tighten the gas cap, and the light should go out.

8,000 feet. (If it's too cold or hot outside and testing takes place at high altitude, the P1000 won't go away.)

Fuel tank level should be between half and three-fourths full (three-fourths full is better for testing).

Check the scan tool for any pending DTCs. Any pending codes must be cleared and/or repaired before the P1000 will clear. After repairs, clear any trouble codes before proceeding.

Drive Cycle Procedure:

**NOTE: The first two steps allow the PCM to go into Monitor Entry so it can begin testing. From this point on, all of the monitors can be tested, or you can test only specific monitors for the purposes of verifying repairs or turning off the MIL. It doesn't matter what order these monitor tests and their drive cycle criteria are performed. However, to keep the PCM from setting a P1000 DTC, the EVAP monitor test should be run before the other monitors.

Connect the scan tool to the DLC and turn the ignition key to ON, but don't start the engine. Cycle the ignition key off, then on again. Clear any DTCs that are present.

Start the engine *without* turning the key to the OFF position and let the engine idle for 15 seconds. (The key was left in the ON position from Step 1.) Drive at 40 miles per hour until the ECT sensor shows at least 170 degrees Fahrenheit.

To initiate the EVAP monitor, the throttle position (as viewed on a scanner's PID display) needs to be between 40 and 60 percent and the fuel level sensor should read between 15 and 85 percent. (This is why the gas tank should be half to three-fourths full). Avoid practicing for an autocross event, as making sharp turns will slosh the fuel around in the tank, making lots of fuel vapors and possibly causing the monitor to fail the EVAP system test unnecessarily. To run the monitor, cruise at 45 to 65 miles per hour for 10 minutes. The EVAP monitor tests the EVAP system's ability to route vapors to the engine where they can be burned. It also tests for any leaks in the system.

To run the heated exhaust gas oxygen, or HEGO monitor, cruise (steady driving) at 40 miles per hour for up to four minutes. This monitor tests the internal oxygen sensor's electrical heaters. The monitor will probably run when performing Step 2. This step should be used when only the HEGO monitor needs to be run.

To run the catalyst monitor, drive in stop-and-go traffic conditions. Make sure to include five different constant cruise speeds that range between 25 and 45 miles per hour during a 10-minute driving period. This monitor tests the catalytic converter's ability to store and release oxygen, and thus its effectiveness at controlling exhaust emissions.

To run the EGR monitor, stop the vehicle and then accelerate to 45 miles per hour at half to three-quarters throttle. Repeat this sequence three more times. Use your scan tool to determine throttle opening by watching the throttle position PID. The EGR monitor tests the EGR system's ability to route exhaust gases into the intake manifold.

To run the secondary air (SEC AIR) monitor, stop the vehicle and let the engine idle with the transmission in drive (neutral if you have a manual transmission) for two minutes. The secondary air monitor tests the air pump and hoses for proper operation and leaks. Not all vehicles have this system. In addition to running the SEC AIR monitor, this test starts the comprehensive components monitor (CCM) for the idle speed control test.

To run the misfire monitor, stop the vehicle and then accelerate to 65 miles per hour. Decelerate (close the throttle but don't step on the brakes) until a speed of 40 miles per hour is reached. Stop the vehicle and repeat this procedure two more times. The misfire monitor tests the ignition system's ability to fire the spark plugs under an engine load.

Chrysler Drive Cycle

Chrysler has specific drive cycles for their catalyst, EGR, oxygen sensor, oxygen sensor heater(s) and EVAP monitors. All other monitors use a different drive cycle. All Chrysler monitors, or tests, are organized and prioritized by the Diagnostic Task Manager software within the PCM. Monitors are run under specific operating conditions called enabling criteria. In particular, the EGR monitor will not run if conflicts with other monitors or if DTCs exist; this makes running it somewhat difficult. The following are the drive cycles for each monitor.

Chrysler uses a number of system-specific OBD-II drive cycles. Be sure to use the right one, and be sure to follow its proper operating procedures.

All Monitor Drive Cycle—Preconditioning requirements include: the MIL must be off and the engine should be cold (no temperature is defined). Warm up the engine for 5 minutes until closed loop is achieved (monitor your scan tool for closed loop status). Drive the vehicle with a steady throttle between 40 to 60 miles per hour for 8 minutes. Stop and let the engine idle for 3 minutes. Drive the vehicle again (steady throttle) above 20 miles per hour for 2 minutes. Stop the car and turn the ignition key off for 10 minutes (this runs the O2 sensor heater monitor).

NOTE** The preceding drive cycle is limited and may not run all the monitors on all Chrysler vehicles. Better results can be obtained by running the monitor-specific drive cycles that follow. A service manual should be consulted for specific vehicle drive cycles.

2004 Chrysler Town & Country: DTC P0480

This Ford Motor Company's relay and fuse center is typical of those found on vehicles from a variety of manufacturers. Many of these relays are controlled by the vehicle's powertrain control module, or PCM. The centralized location of the relays makes diagnosis somewhat easier than it was in the past, when relays and fuses were located under the dash. *Courtesy SCM Hotline*

The following is an instance where the OBD-II diagnostic trouble code led a technician to the right place but did not specifically identify the problem. Code P0480 (Low-Speed Fan Relay Control Circuit) was retrieved by a scan tool on a 2004 Chrysler Town & Country 3.8L minivan. With the right scan tool, and Chrysler's onboard diagnostics, a technician can place the scanner into actuator test mode (ATM), where components like fuel injectors, ignition coils, solenoids, and relays can be operated through the scan tool. In this example, the ATM test that commands the low-speed cooling fan relay to turn the fan on failed. The fan did not come on, and the relay didn't "click" (clicking is an indication that the relay is operational).

The first step in the diagnostic process was to determine why the relay didn't work. In general, automotive computers switch the ground side (on or off) of the relays they control. However, in this case, the powertrain control module, or PCM, controls the power side of the relay. (This was determined by using a vehicle wiring diagram and tracing all the wires connected to the low-speed cooling fan relay.)

By connecting a test light's alligator clip to battery ground, touching the pointy end to the power wire at the relay, and performing the ATM test again, it was confirmed that the PCM was indeed sending a power signal to the relay. When a jumper wire was used as temporary ground wire at the relay and the ATM test was performed a third time, the cooling fan came on, indicating a bad relay ground wire. A vehicle component locator listed a ground junction connector near the left headlight, and sure enough, that's where the bad ground was eventually found. For more information on finding bad grounds and on how relays operate, see *How to Diagnose and Repair Automotive Electrical Systems,* published by Motorbooks.

Catalyst Monitor Drive Cycle—Preconditioning requirements include: the MIL must be off, no DTCs may be present, fuel level should be between 15 and 85 percent full (half to three-quarters in the tank), ECT should be above 70 degrees Fahrenheit, and the engine must have been running at least 90 seconds at an rpm between 1,350 and 1,900. Let the engine idle for five minutes until it reaches closed loop status (monitor closed loop stats on a scan tool). Now drive the vehicle at a steady speed between 30 and 45 miles per hour for two minutes. The catalyst monitor should run.

EGR Monitor Drive Cycle—There are two different drive cycles for the EGR monitor, one for vacuum-controlled EGR systems and another for electronically controlled systems.

The EGR (vacuum controlled) drive cycle has only one preconditioning requirement—the MIL must be off. Start the engine and let it idle for five minutes until it reaches closed loop status (watch a scan tool to monitor closed loop operation). Drive the vehicle at a steady speed between 40 and 60 miles per hour for eight minutes. The EGR monitor should run.

The EGR (electronically controlled) drive cycle has six preconditioning requirements: the MIL must be off, the MAP sensor must read between 0 and 60 kPa, (kilopascals), the ECT should be above 180 degrees Fahrenheit, no pending misfire DTCs should be present (check these PIDs with a scan tool), and the air conditioning should be turned off. Start the engine and let it idle for five minutes until it reaches closed loop status (watch a scan tool to monitor closed loop operation). Drive the vehicle using a steady throttle between speeds of 40 to 60 miles per hour for two minutes. Let the engine idle for three minutes. The EGR monitor should run.

NOTE** The EGR monitor determines EGR flow based on a change in idle quality (when the EGR valve is open at idle, engine idle will be rough). Any conditions that cause an unstable idle (such as running the AC, turning the steering wheel, or having the cooling fan(s) cycle on and off) may prevent the EGR monitor from completing.

EVAP Monitor Drive Cycle—There are two different drive cycles for the EVAP monitor. One drive cycle is for the standard type of EVAP system and another for systems that use a leak-detection pump.

The EVAP (standard type) monitor has three precondition requirements: the MIL must be off, no EVAP-related DTCs should be present, and the fuel level should be between half-full and full. The EVAP monitor may still run if the fuel level is between 15 and 85 percent (check fuel level sensor on scanner for this information). Start the engine and let it idle for five minutes until it reaches closed loop status (watch a scan tool to monitor closed loop operation). Drive with a steady throttle at speeds between 30 and 45 miles per hour for two minutes. The EVAP monitor should run.

Preconditions for the leak-detection pump EVAP system are the same as for the standard type. The vehicle must be cold (sitting overnight or at least eight hours). Start the engine and let it idle for four minutes. Drive in simulated stop-and-go traffic conditions for five minutes (use smooth accelerations and decelerations). Stop and let the engine idle for four minutes. The EVAP monitor should run.

Oxygen Sensor Monitor Drive Cycle—The only precondition requirement is that the MIL should be off. Start the engine and let it idle for five minutes until it reaches closed loop status (watch a scan tool to monitor closed loop operation). Using a steady throttle, drive the vehicle at speeds above 25 miles per hour for two minutes. Stop and let the engine idle for 30 seconds. Accelerate smoothly to a speed between 30 and 40 miles per hour. Repeat the last step four more times. The oxygen sensor monitor should run.

Oxygen Sensor Heater Monitor Drive Cycle—There is only one precondition for this monitor: the MIL should be off. Start the engine and let it idle for 5 minutes until it reaches closed loop status (watch a scan tool to monitor closed loop operation). Shut the engine off for 10 minutes. The oxygen sensor heater monitor should run.

European Drive Cycle

Since 2002, all automobile imports from Europe must meet the same standards as OBD-II vehicles made in the United States. These imports use a system called EOBD, which is very similar to an OBD-II system. The following is a generic EOBD drive cycle that should work for verifying repairs and turning off the MIL for most European imports.

The PCM will start a drive cycle as soon as the engine is started (from either hot or cold conditions) and ends the drive cycle when the engine is turned off. To meet all the conditions that will allow all monitors to run and have a completed status, the following steps should be taken:

The European, or EOBD, drive cycle starts as soon as the engine is started, whether hot or cold, and ends when the engine is shut off. Just like the domestic manufacturers' drive cycles, different manufacturers may use slightly different procedures. A service manual will provide details for testing specific brands and models of European imports.

Cold start. For the PCM to recognize a cold start the coolant temperature must be below 50 degrees Celsius (122 degrees Fahrenheit) and within 6 degrees Celsius (42 degrees Fahrenheit) of the ambient air temperature when the engine is started.

Idle. The engine should idle for two to three minutes with several electrical loads turned on, such as the rear screen defogger or heater blower fan. This will allow the oxygen sensor heater, misfire, and fuel trim monitors to run.

Accelerate. Accelerate the vehicle to a speed between 50 and 60 miles per hour (that's 80 to 96 kilometers per hour for the metrically inclined). The misfire and fuel trim diagnostics will be performed.

The vehicle speed should now be held steady between 50 and 60 miles per hour for about three minutes. The PCM will run the O2 response, misfire, and fuel trim monitors.

At the end of this three-minute period (Step 4) the accelerator pedal should be released and the vehicle allowed to decelerate to 20 miles per hour without changing gears or using the brakes or clutch. Fuel trim diagnostics will again be run.

Accelerate the vehicle again to between 50 to 60 miles per hour. The misfire and fuel trim diagnostics will be performed again as in Step 3.

Hold the vehicle's speed steady at between 50 and 60 miles per hour for about five minutes. The catalyst monitor will run. If the catalyst efficiency is low (the converter is worn out after many miles) or if the battery has been disconnected, it may take up to five complete drive cycles for the PCM to determine the effectiveness of the catalyst.

Finally, the vehicle should be decelerated, as per Step 5, and then stopped.

Common Diagnostic Trouble Codes and Their Causes

The following are some of the more common generic OBD-II diagnostic trouble codes and their possible causes. After reading any stored DTC, but before replacing any parts, always check for manufacturer technical service bulletins (TSBs). These documents provide a wealth of information and may have specific procedures for the trouble codes that are present on the vehicle being worked on. They also may indicate that a software update is required for the vehicle's PCM or a PCM replacement as the solution for various drivability issues, or as a way to keep the MIL and false DTCs from coming back to haunt the vehicle owner. If more information regarding generic and manufacturer-specific DTCs is needed, they can be found in a vehicle's service manual, or by surfing the Internet. Also, there are generic OBD-II code definitions in the appendix of this book.

P0455—Evaporative Emission System Leak Detected (Gross Leak/No Flow)

This code indicates a "gross" (large) leak in the EVAP system. Did you remember to tighten the gas cap the last time you filled the tank? If the gas cap was left off, or just loose, a PO455 DTC could be set by the PCM. Also, if an aftermarket gas cap is used it may not seal as well as the original part, causing this code to be set. Other causes for large leaks include a broken canister vent valve, disconnected vacuum or EVAP system vent hoses, or a damaged vapor canister.

P0442—Evaporative Emission System Leak Detected (Small Leak)

The same problems that caused a P0455 can also cause this DTC to be set. Again, look for any EVAP system hoses that have cracked from old age, come loose, or been disconnected. Also, carefully inspect the gas cap for rubber O-rings or gaskets that are damaged or missing. Just as with the P0455 you are looking for a leak, only this time the leak will be small.

P0440—Evaporative Emission System

This code indicates that the EVAP system is not functioning, but not necessarily leaking. There are 20 generic OBD-II EVAP-related codes and many more manufacturer-specific EVAP trouble codes. The presence of any of these DTCs may provide additional specific information as to the cause of the P0440 code. Components to check include gas cap, EVAP vent hoses, vacuum lines, canister vent valve, purge solenoid, and fuel lines.

P0420—Catalyst System Efficiency Below Threshold, Bank 1
P0430—Catalyst System Efficiency Below Threshold, Bank 2

The PCM will set these codes when one of the rear (located after the converter) oxygen sensor output voltages varies between 0.2 and 0.6 volt. This indicates that the catalyst is no longer reducing exhaust emissions sufficiently. In addition to a worn-out catalyst, there are other causes for these codes including: oxygen sensors, or their circuits, that are malfunctioning (both up- and downstream O2 sensors); an exhaust system leak; or other exhaust damage. A software update for the PCM or new PCM should also be looked at as a possibility for these codes, as some manufacturers have made the catalyst efficiency monitor (and the codes it produces for failures) a bit too sensitive, causing the P0420/30 codes to be set even when the catalytic converter is working properly.

P0401—Exhaust Gas Recirculation Flow Insufficient Detected

Common causes of incorrect EGR flow include a damaged or blocked EGR vacuum sensing hose, clogged EGR valve or exhaust hose, EGR passageway plugged due to excessive carbon buildup, defective EGR vacuum solenoid, no vacuum signal to the solenoid, or no command from the PCM to the EGR vacuum control or electronically operated EGR valve. In addition, various EGR-related codes can be caused by low exhaust backpressure from some high-flow exhaust systems installed by the owner.

Auto manufacturers design a certain amount of exhaust back pressure to be present in their exhaust systems. Some EGR systems use a differential pressure sensor, sometimes referred to as differential pressure feedback exhaust (DPFE), to regulate EGR flow into the intake manifold during EGR operation. When an aftermarket, high-flow exhaust is installed, the DPFE never has enough exhaust backpressure to operate the EGR system correctly and the PCM may set EGR-related DTCs. This picture shows a larger-than-stock exhaust system that could cause an EGR code to be set. The big wing keeps this hot rod stable at speeds over 200 miles per hour—at least that's what the owner told me.

P0300—Random Multiple Misfire Detected

Random or general ignition misfire can commonly be caused by primary and secondary ignition electrical problems. On vehicles that use a distributor, excessive cap and rotor wear will affect the entire ignition system, causing the random multiple misfires. Also, any malfunctions with fuel delivery, air metering, excessive EGR, or an intake restriction can cause a cylinder(s) to misfire. The mechanical condition of the engine should not be overlooked, as leaking valves or compression problems can also cause misfires.

There are several seemingly unrelated causes for the P0300 random misfire code. If the vehicle was used for off-road activities (Jeeps, trucks, and SUVs) and the tires and wheels picked up lots of mud, they may become unbalanced. Also, if one of the wheel weights fell off, a tire can become out of balance. These out-of-balance conditions cause vibration in the driveline, which is transmitted to the engine's crankshaft, causing it to speed up and slow down. The PCM is not smart enough to tell the difference between a real ignition misfire and vibration from an out-of-balance wheel, or wheels, and could set a random misfire code. Also, imbalances in the driveshaft can cause the same false DTC problems.

P0301, 2, 3, 4, 5, 6, 7, 8—Cylinder X Misfire Detected

Misfire for cylinder X (X = cylinder number) is commonly caused by primary and secondary ignition faults, incorrect fuel delivery, air metering, excessive EGR, intake restriction, or engine mechanical problems. In addition, a vacuum leak at one of the port fuel injectors or intake runner may cause one or more cylinders to misfire due to a too-lean fuel mixture in that cylinder, caused by the vacuum leak.

P0171 (Bank 1) & P0174 (Bank 2)—System Too Lean

A system-too-lean DTC for only one bank of cylinders can be caused by an intake port or fuel injector vacuum leak on one side of the engine. Also, a large vacuum leak may cause the PCM to set both codes (P0171 and P0174). In addition, incorrect MAP or MAF sensor readings can cause lean-fuel-mixture-related DTCs to be set. Low fuel pressure from an electric fuel pump circuit with high resistance or one with a weak or tired fuel pump will also cause overly lean fuel mixtures that may result in these types of DTCs.

P0172 (Bank 1) & P0175 (Bank 2)—System Too Rich

Causes for a rich condition DTC from only one bank of cylinders include poor fuel injector spray pattern (bad injector), malfunctioning oxygen sensor for that bank of cylinders, and incorrect air metering from an MAF or MAP sensor. A leak in the fuel pressure regulator diaphragm could cause both codes to be set because this condition will cause all the engine's cylinders to be too rich. In addition, malfunctioning oxygen sensors can cause too much fuel to be added to the engine, and, thus, set these diagnostic trouble codes.

P0106—Manifold Absolute Pressure/Barometric Pressure Circuit Range/Performance

A manifold absolute pressure (MAP) sensor malfunction can cause a false signal to be sent to the PCM that is out of range with what the PCM expects. This can be caused by a cracked, leaking, or loose MAP vacuum hose. (Some MAP sensors are connected directly to the engine's intake manifold and don't use a vacuum hose, but could leak nonetheless.) Also, a bad electrical connection at the MAP sensor or open/short in the wiring harness between the MAP and the PCM may also cause this and other related MAP sensor codes including: P0105—MAP circuit, P0107—MAP circuit low, P0108—MAP circuit high, and P0109—MAP circuit intermittent. When testing a MAP sensor and comparing its values with a mechanical vacuum gauge or reading displayed on a scan tool, make sure that the scanner and gauge are using the same units of measurement. MAP vacuum levels can be read in inches of mercury (In-Hg), kilopascals (kPa), or pounds per square inch (psi).

P0115—Engine Coolant Temperature Circuit

This DTC can be set by the PCM for a malfunction in the coolant sensor circuit. This could be caused by something as simple as the coolant sensor that is unplugged or one with a poor electrical connection. In addition, the PCM may sense that the coolant temperature is too high or low for an extended period of time and set this DTC.

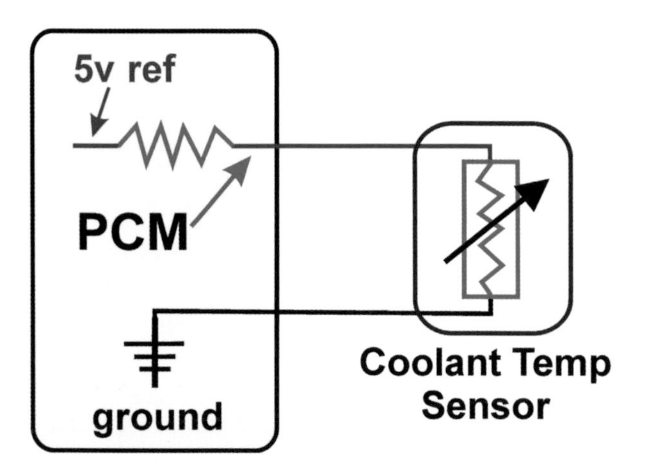

Fig. 7-3—The PCM in an OBD-II system reads the coolant temperature sensor's voltage at a point on the temp sensor's signal wire located just before an internal resistor (blue arrow). This variable voltage signal is interpreted by the PCM as engine temperature—the higher the engine's temperature, the lower the sensor voltage. (Many vehicles will read less than 1.0 volt when the engine is at normal operating temperature.) On a cold engine that has sat over night, the coolant temperature sensor's voltage will be around 3.5 volts. A scan tool or digital voltmeter can be used to read coolant temperature sensor outputs. More information on how these sensors operate and how to diagnose them can be found in the book *Automotive Electrical Systems, Diagnosis and Repair* by Tracy A. Martin.

P0116—Engine Coolant Temperature Circuit Range/Performance

The PCM can set this DTC because the coolant sensor's voltage signal does not agree with other sensor input data. For example, if the engine has not been started for several hours or overnight, and the air temperature sensor reads 50 degrees Fahrenheit, the coolant sensor should have a similar (if not the same) reading. If the coolant temperature sensor (CTS) reads 120 degrees Fahrenheit, or some other value that is out of line with the air temperature sensor, the PCM could set this range/performance code. In addition, if there is a problem with the vehicle's thermostat (stuck open for instance) the engine may take too long to reach normal operating temperature. The PCM knows how long the engine has been running since it was started and expects to see the CTS input rise at a predetermined rate. If the thermostat was stuck open, the engine will take longer to warm up and the PCM may set this code.

P0117—Engine Coolant Temperature Circuit Low

The engine coolant sensor may be unplugged or its signal wire is broken somewhere between the sensor and the PCM. A scan tool may show the CTS temperature as very cold, 0 degrees or below. A bad coolant sensor can also cause the CTS to read very low temperatures and the PCM could set this code.

P0118—Engine Coolant Temperature Circuit High

The engine coolant sensor may have a short in its signal wire between the sensor and the PCM. This causes the signal to the PCM to indicate a very high temperature usually above 250 degrees Fahrenheit. A bad coolant sensor can also cause this high temperature condition and cause the PCM to set this code.

P0119—Engine Coolant Temperature Circuit Intermittent

This DTC is usually caused by a poor or loose connection at the coolant temperature sensor. It could also be caused by a loose connector at the PCM or in a wiring harness. A bad coolant temperature sensor can also cause an intermittent signal as well setting this DTC.

P0011 (12, 14, 21, 23, 24 and 25)—"A" or "B" Camshaft Position, Timing Over Retarded/Advanced, Bank 1 or 2

Any related camshaft or crankshaft DTCs can often be caused by timing belt issues. If the timing belt was installed incorrectly (off by one tooth or more), a no-spark condition may be caused as well as the presence of these codes. The no-spark condition is caused because the PCM is trying to prevent an intake backfire that could blow the air cleaner off. An intake manifold backfire can be caused by retarded camshaft timing. Not all OBD-II systems use the strategy.

In addition, if the timing belt on the vehicle has more than the recommended miles on it and needs to be changed, it can stretch just enough to cause one of these

While in the process of doing research for this book the author takes a moment to collect his thoughts. The DLC is right where the service manual said it would be—next to the choke pull knob. But when he tries to read the DTCs, his scan tool says "No Communication, PCM Not Found." He wonders why the scanner can't find the PCM? Is it lost? Maybe he forgot to turn the ignition switch to ON, or maybe this is one of those cars that uses that new CAN protocol and the scanner can't read the PIDs from the PCM. *Courtesy Elwoods Auto Exchange*

"Timing Over Retarded" or "Timing Over Advanced" related trouble codes. Another possible cause for a slightly retarded or advanced camshaft is a brand-new timing belt. Some aftermarket timing belts are a couple of degrees short and need a few miles to stretch out. However, if the new belt also causes a no-spark condition, the engine will not run and consequently the timing belt will never stretch. The best solution is to use a factory timing belt.

P0016 (17, 18, and 19)—Crankshaft Position to Camshaft Position Correlation, Bank 1 or 2, Sensor A or B

These trouble codes are similar to the pervious DTCs in that they relate to camshaft timing issues. More specifically these codes relate to crankshaft vs. camshaft timing, or correlation. The PCM is looking at the camshaft and crankshaft sensors and determining if the cam and crank are in the correct timing relation to each other. Take a close look at the timing belt. Is it past its useful service life and needs to be replaced as a maintenance item? Or is it brand-new and is slightly too short and needs to be stretched by diving the vehicle for a few miles? Also, these codes could cause a no-spark condition. Because of the oversensitivity of some OBD-II systems, it is often a better choice to use a timing belt from the vehicle manufacturer instead of the automotive aftermarket to avoid these types of DTCs.

This chapter brings this book to a close. I hope that those of you who have stuck it out and read the whole book have learned something about automobile emissions, electronic fuel injection, and onboard diagnostics. I learned a great deal in doing the research for the book and in writing it. If you have any comments about the book please contact me via my website: www.TracyAMartin.com. Looking forward to hearing from you soon.

Tracy Martin . . .

SOURCES

I would like to thank the following companies for help with the images and information contained in this book. Their kind assistance made this a better book both from technical and artistic perspectives. All of these automotive dealers and aftermarket manufacturers offer great products and services for professional and do-it-yourself technicians alike. Contact information is listed for each company—check out their websites.

Actron, an innovator in quality automotive diagnostic tools and accessories, continues its 40-plus-year tradition as a leading manufacturer of cost-effective, high-performance diagnostic and test equipment. From scan tools and code readers, to multimeters, battery testers, and timing lights, Actron has an automotive diagnostic tool for every user level—from the DIYer to the professional technician. Actron's innovative products are the tools of choice for do-it-yourselfers.

Actron
800-228-7667
www.actron.com

AutoXray is a leading provider of effective, affordable diagnostic equipment to both amateur and professional automotive technicians. The company features its comprehensive line of EZ-SCAN scanners and code readers. Product accessories include Windows-based software to interface EZ-SCAN with PCs and the Internet, and battery testers that plug into EZ-SCAN diagnostic scanners to measure battery condition, voltage, and available power.

AutoXray
800-595-9729
www.autoxray.com

OTC is a major manufacturer and supplier of vehicle electronic diagnostic instruments, automotive fuel system maintenance equipment, special-service tools, general-purpose tools, pullers, heavy-duty tools, shop equipment, and hydraulic components. The company markets its products through a global network of authorized aftermarket distributors, and serves a large number of car, truck, agricultural implement, construction, and marine manufacturers on an OEM basis.

OTC/SPX Corporation
800-533-5338
www.otctools.com

S.C.M. Hotline

With the ever-escalating price of automotive parts (especially the electrical ones) you can't afford to fix your car or truck with the "try a known good part" repair method. Auto parts stores and new car dealerships will not take back any part that is electrical in nature. You need to know for sure what parts, associated wiring, and onboard computers are causing an emission or drivability problem before reaching for your wallet. Some quality diagnostic help from S.C.M. Hotline goes a long way to fixing your car right the first time and can save you considerable time and money as well.

S.C.M. Hotline, located in Garden Grove, California, currently supports a large number of professional technicians with quick access to technical data from the Hotline's in-house database. Vehicle coverage includes both foreign and domestic cars and light trucks (gasoline and diesel) from 1964 to present. The automotive systems covered include: engine performance, OBD-I/OBD-II diagnostics, climate control (HVAC), ABS braking systems, computer-controlled transmissions, supplemental restraint systems (airbags), and electronic body control systems.

When you place a call to the hotline you will be connected with a technician immediately—no callbacks at a later time. Your personal technician will talk you through the diagnosis steps required to solve your automotive problem. The average call time for 89 percent of customer inquiries is two 7-minute phone calls.

The staff automotive repair experts are Automotive Service Excellence (ASE) certified Master Technicians with years of experience in the automotive industry. They have access to all the latest OEM factory information, technical service bulletins, wiring diagrams, repair procedures, component locators, and other service-related information.

Please call for details on purchasing diagnostic time with the hotline. The diagnostic phone service is open from 6:00 a.m. to 5:00 p.m. Pacific Standard Time, Monday through Friday. S.C.M. Hotline offers you a place to turn to for help with all your automotive diagnostic and repair questions.

S.C.M. Hotline
800-847-9454
www.AutoHotLineUsa.com
Email: SCM@autohotlineusa.com

ScanTool.net

Inexpensive solutions for OBD-II diagnostics

ScanTool.net, LLC

ScanTool.net, LLC, is based in Phoenix, Arizona, and is the world leader in the development and manufacturing of inexpensive solutions for onboard diagnostics (OBD-II). The company was founded in 2002 with a single goal in mind: to build an affordable, yet sophisticated, personal scan tool—which was realized in the ElmScan™ family of scan tools.

That same year, the company released the first open source software for onboard diagnostics. This open architecture approach, coupled with low-cost OBD-II hardware, enabled car enthusiasts and software companies the world over to develop diagnostic software for the ElmScan™ scan tools. Over a dozen free and commercial diagnostic programs are available, designed for a number of different applications: engine diagnostics, performance tuning, digital dashboard, data logging, and fleet management. A wide variety of hardware platforms (PCs, PDAs, Internet Tablets) and operating systems (Windows, DOS, Linux, and many others) are supported.

Today, ScanTool.net offers a wide range of inexpensive PC-based scan tools, covering all existing and future OBD-II-compliant vehicles. The scan tools can be connected to the PC using serial or USB cables, or even wirelessly via a Bluetooth connection.

ScanTool.net, LLC, places a great emphasis on customer service and technical support. They stand behind their products, and their customers enjoy a generous "no questions asked" return policy, and the best warranty in the business.

ScanTool.net, LLC
1819 W. Rose Garden Ln, Ste 3 Phoenix, AZ 85027
(866) 923-1870
www.ScanTool.net
E-mail: sales@scantool.net

EASE Diagnostics

EASE Diagnostics is the industry leader in PC-based automotive and information products with over 12 years of success, along with numerous industry awards for innovative product design in the highly competitive professional diagnostic market.

The EASE suite of diagnostic and information products is available to the consumer using the same care and techniques as used in their professional products. Whether you have an in-car PC, PDA, laptop, or desktop, you can access the most useful and powerful suite of automotive diagnostic and information offerings ever available.

EASE Diagnostics
Scott Technology Park
RR1, Box 285
Olyphant, PA 18447
PH: (888) 366-3273
www.obd2.com

AutoTap OBD-II Diagnostic Scanner by B&B Electronics Manufacturing Company

B&B Electronics Manufacturing Company designs and manufactures diagnostic and telematics solutions for fleet managers, performance enthusiasts, and the do-it-yourself mechanics. They sell throughout the world with primary markets being North America, Europe, Middle East, and Africa.

The U.S. headquarters in Ottawa, Illinois, houses engineering, manufacturing, service, and warehousing. The European headquarters, which provides service, support, and warehousing, is located in County Galway, Ireland.

In addition to manufacturing products, B&B is a value-added distributor for other quality vendors. This gives B&B the depth to solve most any data communications problem and be a one-stop source for data communications products.

The B&B culture is one of absolute devotion to the customer, always acting with integrity and honesty, having fun, producing outstanding quality products and service, and providing their customers with the highest value.

B&B Electronics
707 Dayton Road
PO Box 1040
Ottawa, IL 61350
815-433-5100
www.autotap.com

Younger Toyota

Since 1976 the Younger family has been serving its area with the best automotive experience available. Being a family-owned operation allows Younger to take a caring attitude toward its customers for long-term relationships. Younger Toyota is a nine-time winner of Toyota's prestigious "Presidents" award. Not only has Younger performed well with Toyota's highest customer service award, it has won Sales (15 years running), Service and Parts (12 years running) Excellence awards. No other store in the Maryland area carries these credentials! What this means to the customer is an operation built around customer service with a dealer who really takes you seriously.

Younger Toyota
1935 Dual Highway
Hagerstown, MD 21740
301-733-2300
www.youngertoyota.com

Keplinger's Automotive Center

In February 1968 my career in the automotive repair industry began when I opened Williamsport Shell, a gas station with two service bays. Soon after opening, I realized that if I wanted to succeed as an independent repair shop, I needed additional training from the basics of what I already knew, and so began a 40-year journey to learn how to repair automobiles—a journey that has no end. Allen Test Products, Catonsville Community College, and numerous after-hour automotive seminars provided just some of the training for me and the technicians who worked for me.

I spent 27 years running the service and repair stations for Shell and Exxon. During that time I was twice elected to represent Baltimore-area Exxon dealers at Exxon corporate headquarters in Houston Texas. In 1994 I opened a repair-only shop in a leased building and ran that business for five years.

Due to the success of the business, and my expanding customer base, I couldn't take care of all the customers that wanted service in the current building, so in 1999 I built a 10-bay service center—Keplinger's Automotive Center, located in Hagerstown, Maryland. We work on all makes and models of cars and trucks and have a total of six technicians, including two of my sons. If you're in the area, and need service for your vehicle, please stop by.

Larry C. Keplinger

Keplinger's Automotive Center
10218 Sharpsburg Pike
Hagerstown, MD 21740
301-733-0760

This 1953 Willys Arrow Lark has been parked here since 1953. Elwood Auto Exchange has lots of great old cars and trucks.

Level Five Graphics, Inc.

Level Five Graphics, Inc., is a graphic design, illustration, and photography firm that provides services in support of the advertising, marketing, and publishing industries, as well as business-to-business work such as corporate identity. If you need printed materials, Level Five can produce it for you, from start to finish. Please visit www.levelfive.com, or call 301-933-3906 to discuss your project. What can Level Five do for you?

Elwoods Auto Exchange

Elwoods Auto Exchange is located in Smithsburg, Maryland, on Route 64. Established in 1950, Elwood Grimm's first junked car is this 1953 Willys Aero Lark and it has been parked right outside the old farmhouse ever since they opened for business. Elwoods has cars and trucks (even a few motorcycles) from the 1940s to present in various states of salvage. Nestled in rural Maryland farmland, the yard covers 53 rolling acres packed full of old and not-so-old cars and trucks.

According to William Smith, the company historian, Elwoods is famous, or infamous depending on whom you talk to. It seems that President Lyndon Johnson and the first lady, Ladybird Johnson, were on their way from Camp David, located in the Catoctin Mountains a few miles away, to church in Hagerstown. As the presidential motorcade passed the junkyard, Ladybird asked that they stop. She got out of the limousine and marched up the dirt road to Old Man Grimm's office and gave him a piece of her mind regarding the unsightliness of the wrecking yard and how it ruined the beautiful landscape. This was in 1963, and in 1965, the Highway Beautification Act was passed by Congress. This legislation authorized federal funding for the removal of billboards and the cleaning up of junkyards along interstate and other highways. So we have Elwoods to thank for having wrecking yards located away from (or at least hidden from) major highways.

Elwoods was a great place to take many of the photographs for this book. If I needed a photo of a crank sensor from a late-model Chevy, I just walked along the rows of cars until I found one with the front end ripped off and bent down, and got the shot I needed. Images like this would be extremely difficult to get from a running automobile as there is no way to even get a camera close to some of the components I took pictures of. I like to thank Elwoods for letting me photograph their great selection of cars and trucks. They are open seven days a week, from 8 a.m. to 5 p.m.

Elwoods Auto Exchange
21411 Jefferson Boulevard
Smithsburg, MD 21783
Phone: 301-739-7159
www.elwoodsautoexchange.com

APPENDIX

Domestic OBD-I and OBD II Applications

This table shows the various onboard engine management systems used by domestic auto makers and emphasizes how different they all were until OBD-II in 1996.

GM Onboard Diagnostics

System	Years Used	Description
OBD-I Control Module	1981–1995	Most vehicles used the 12-pin ALDL (Assembly Line Data Link) located under the dash on the driver side. Some 1994-1995 vehicles used the 16-pin OBD-II (J1962) data link connector (DLC) but use the Historical application software. Refer to the vehicle's Vehicle Emission Control Information label.
OBD-II Control Module	1994*–Present	Complies with OBD-II regulations and uses the J1962 DLC.

** OBD-II system is used on certain 1994–1995 vehicles equipped with a 2.2L, 2.3L, 3.8L, 4.3L, or 5.7L engines.*

Ford Onboard Diagnostics

System	Long Name	Years Used	Description
MCU	Microprocessor Control Unit	1980–1991	Used in police vehicles containing carbureted engines. Uses the MCU DLC.
EEC-IV	Electronic Engine Control Fourth generation	1984–1995	Most Ford vehicles equipped with North American engines. Uses the EEC-IV DLC.
MECS	Mazda Electronic Control System	1988–1995	Vehicles equipped with Mazda-sourced engines. Uses MECS 6-pin and 17-pin DLCs.
EEC-V	Electronic Engine Control Fifth generation	1994*–present	Complies with OBD-II regulations and uses the OBD-II J1962 DLC.
PTEC	Powertrain Electronic Controller	2000–present	Complies with OBD-II regulations and uses the OBD-II J1962 DLC.

** EEC-V OBD-II system used in 1994–1995 vehicles equipped with a 3.8L or 4.6L engine.*

Chrysler Onboard Diagnostics

System	Long Name	Years Used	Description
SMEC	Single Module	1989–1990 Engine Controller	Used a 6-pin Serial Communication Interface (SCI) DLC and has bidirectional capability.
SBEC	Single Board Engine Controller	1989–1995	Used two types of DLCs; a 6-pin (SCI) and 6-pin LH series.
OBD-II	Powertrain Control Module	1995–present	Uses OBD-II J1962 DLC connector.
JTEC	Jeep/Truck Engine Controller	1996–present	JTEC system is used on light-duty trucks and Jeeps.

These tables provide a partial list of trouble code definitions for domestic automobile manufacturers. Chapter 1 covers how to retrieve OBD-1 codes from all three automakers.

OBD-I Domestic Code List

General Motors OBD-I Code List

Code	Definition
Code 12	No RPM Signal
Code 13	O_2 Sensor
Code 14	CTS Shorted
Code 15	CTS Open
Code 16	High Volts
Code 21	TPS High
Code 22	TPS Low
Code 23	MAT Sensor
Code 23	M/C Sol. (carb)
Code 24	VSS
Code 25	MAT
Code 26	Quad Driver
Code 27	Trans Switches
Code 28	Trans Switches
Code 29	Trans Switches
Code 31	P/N Switch
Code 32	EGR (EFI)
Code 32	BARO (carb)
Code 33	MAP High
Code 33	MAF High (pfi)
Code 34	MAF Low (pfi)
Code 36	Burn Off
Code 36	High Air Flow
Code 38	BOO (quad 4)
Code 39	TCC (quad 4)
Code 40	P/S
Code 41	No Dist Signal
Code 43	ESC
Code 44	O_2 Lean
Code 45	O_2 Rich
Code 46	VATS
Code 47	ECM/BCM Data
Code 51	PROM/MEM CAL
Code 53	EGR (carb)
Code 48	MISSFIRE (VIN C)
Code 52	CALPAC
Code 53	High Battery (fi)
Code 54	Fuel Pump (fi)
Code 55	Ground ECM
Code 62	Gear Switch
Code 63	EGR (3800/quad 4)
Code 64	MAP (pfi)
Code 64	EGR (3800/quad 4)
Code 65	EGR (3.8L & 3800)
Code 65	Injector
Code 66	ECM Reset

Ford Motor Company OBD-I Code List

Code	Definition
Code 11	System Pass
Code 12	Min Idle Speed Too Low
Code 13	Min Idle Speed Too High (PFI)
Code 13	ISC Not Responding (TPI)
Code 14	PIP Erratic
Code 15	No Battery Volts
Code 16	RPM Too Low
Code 17	Idle RPM Dropped on Self-Test
Code 18	Loss of IDM or SPOUT
Code 19	ECA Battery +
Code 21	CTS KOEO
Code 22	MAP/BARO Wrong Off Idle
Code 23	TPS High
Code 25	Knock Sensor
Code 27	VSS
Code 28	Loss of TACH/VAT
Code 31	BVP/PFB Below .2 volt
Code 32	EVP/PFB Below .24 volt
Code 33	EGR Opening Not Detected
Code 34	EVP/PFE Volts Too High
Code 35	EVP/PFE Above Max Voltage
Code 36	Cruise Control
Code 38	Nose Switch
Code 39	TCC/AXOD Trans
Code 41	O_2 Lean
Code 42	O_2 RICH
Code 44	AIR Mgt. Inop.
Code 45	AIR Mgt. Upstream
Code 46	AIR Mgt. Not Bypassed
Code 47	Rich/IAC
Code 48	Injector
Code 51	CTS
Code 52	P/S Press Switch
Code 53	TPS
Code 54	VAT/ACT

Code	Definition
Code 55	Key Power Open
Code 56	VAF
Code 57	PIN Switch
Code 58	Nose Switch
Code 59	3rd/4th Gear Switch
Code 61	CTS Shorted
Code 62	2nd/3rd/4th Gear Switch
Code 63	TPS Low
Code 64	VAT/ACT
Code 65	BATT/O_2
Code 66	VAF
Code 67	Neutral or AC On
Code 68	Nose Switch/Trans. Temp.
Code 69	3rd/4th Switch
Code 71	Nose Switch
Code 72	Power Interrupt/No Goose Test
Code 73	TPS
Code 74	BOO
Code 76	Airflow

Chrysler OBD-I Code List

Code	Definition
Code 11	RPM
Code 13	MAF
Code 15	VSS
Code 12	Memory
Code 14	MAP
Code 16	Low Battery
Code 17	ENG COLD
Code 22	CTS
Code 23	PFI ACT
Code 24	TPS
Code 25	AIS (IAC)
Code 26	Injector Circuit
Code 27	Inj Control
Code 31	Purge Soleniod
Code 32	Pwr Loss Light
Code 32	EGR (truck)
Code 33	A/C Cut Out
Code 34	EGR Solenoid
Code 34	Cruise Control
Code 35	Fan Relay
Code 35	Nose Switch (truck/van)
Code 36	AIR Switch Solenoid (truck)
Code 36	Waste Gate (pfi)
Code 37	BARO Read Solenoid (turbo)
Code 37	TCC (auto trans, truck)
Code 37	Shift Light
Code 37	Lock Up Converter Relay
Code 41	Charging System
Code 42	ASD Relay
Code 46	Battery High
Code 47	Battery Low
Code 51	O_2 Sensor Lean
Code 52	O_2 Sensor Rich
Code 53	Logic/SMEC
Code 54	Logic Module (84 ONLY!)
Code 54	Dist Signal
Code 55	End Codes
Code 61	BARO Read
Code 63	EEPROM Write
Code 88	Start Codes

Vehicle Manufacturer Contact Information for OBD-II

Domestic Vehicles	Web Site Address	Phone Number
General Motors		
Chevrolet	www.chevrolet.com	1-800-551-4123
Pontiac	www.pontiac.com	1-800-551-4123
Oldsmobile	www.oldsmobile.com	1-800-551-4123
Buick	www.buick.com	1-800-551-4123
Cadillac	www.cadillac.com	1-800-333-4CAD
Saturn	www.saturn.com	1-800-553-6000
Ford		
Ford	www.ford.com	1-800-392-3673
Lincoln	www.lincoln.com	1-800-392-3673
Mercury	www.mercury.com	1-800-392-3673
Chrysler		
Chrysler	www.chrysler.com	1-800-348-4696
Dodge	www.dodge.com	1-800-348-4696
Plymouth	Not Available	1-800-348-4696
Eagle	Not Available	1-800-348-4696
European Vehicles		
Audi	www.audi.com	1-800-544-8021
Volkswagon	www.vw.com	1-800-544-8021
BMW	www.bmw.com	1-201-307-4000
MINI	www.mini.com	1-201-307-4000
Jaguar	www.jaguar.com	1-800-4-JAGUAR
Volvo	www.volvo.com	1-800-458-1552
Mercedes-Benz	www.mercedes-benz.com	1-800-367-6372
Land Rover	www.landrover.com	1-800-637-6837
Porsche	www.porsche.com	1-800-PORSCHE
Saab	www.saab.com	1-800-955-9007
Asian Vehicles		
Acura	www.acura.com	1-800-999-1009
Honda	www.honda.com	1-800-999-1009
Lexus	www.lexus.com	1-800-255-3987
Scion	www.scion.com	1.866.70.SCION
Toyota	www.toyota.com	1-800-GO-TOYOTA
Hyundai	www.hyundai.com	1-800-633-5151
Infiniti	www.infiniti.com	1-800-662-6200
Nissan	www.nissanusa.com	1-800-NISSAN1
Kia	www.kia.com	1-800-333-4542
Mazda	www.mazda.com	1-800-222-5500
Daewoo	www.daewoo.com	1-822-759-2114
Subaru	www.subaru.com	1-800-SUBARU3
Isuzu	www.isuzu.com	1-800-255-6727
Geo	Not Available	Not Available
Mitsubishi	www.mitsubishi.com	1-888-MITSU2004
Suzuki	www.suzukiauto.com	1-800-934-0934

OBD-II Automotive Terminology

Term	Definition
ACT	Air Charge Temperature sensor, measures air temperature entering the engine.
AFM	Airflow Meter, mechanically measures volume of air entering the engine.
ALDL	Assembly Line Data Link, GM connector for reading codes engine data. See DLC.
BARO	Barometric Pressure sensor, measures altitude or atmospheric pressure.
BCM	Body Control Module, controls accessories and functions other than the engine and transmission.
BOO	Brake On/Off switch, tells the PCM when the brakes are being applied.
BUS	The electronic circuits (hardware) that connects a scanner to an OBD-II system.
CCM	Comprehensive Component Monitor, one of 12 OBD-II diagnostic monitors or tests.
CO	Carbon Monoxide, an odorless poisonous gas that is a product of incomplete combustion.
CO_2	Carbon dioxide, an inert gas that is a byproduct of the combustion process.
CAN	Controller Area Network, a communication protocol that is mandatory by 2008 for OBD-II.
CAT	Catalytic Converter, mounted in the exhaust to control HC, CO, and NO_X from the exhaust.
CKP	Crankshaft Position sensor, measures the rotational speed and position of the crankshaft.
CMP	Camshaft Position sensor, monitors camshaft position, used for cylinder identification.
CTS	Coolant Temperature Sensor, monitors engine coolant temperature.
DI	Distributor Ignition, uses a conventional distributor with internal electronics.
DIC	Data Information Center, data display located in the dash, center, or overhead console.
DICM	Driver Information Center Module, same function as DIC.
DIM	Dash Integration Module, controls memory seats, mirror tilt steering wheel, radio memory.
DIS	Distributorless Ignition System, uses remote ignition module and coil pairs or coil over plug.
DLC	Data Link Connector. Diagnostic Link Connector, same as ALDL.
DMM	Digital Multimeter, solid state multimeter for measuring volts, amps, ohms, and others.
DPFE	Differential Pressure Feedback EGR, Ford acronym for EGR valve flow sensor.
DTC	Diagnostic Trouble Code, the number(s) stored in the PCM to indicate a malfunction.
DVOM	Digital Volt-Ohmmeter, see DMM above.
ECT	Engine Coolant Temperature sensor, same component as CTS.
EEPROM	Electronically Erasable Programmable Read-Only Memory.
EFE	Early Fuel Evaporator Valve, helps warm up intake manifold with exhaust gas.
EGO	Exhaust Gas Oxygen sensor, measures amount of oxygen in the exhaust flow.
EGR	Exhaust Gas Recirculation, valve that meters exhaust gas into intake manifold to be reburned.
EI	Electronic Ignition, generic description of any distributorless ignition system.
EPA	Environmental Protection Agency, Federal Government.
EVAP	Evaporative System, controls fuel vapors from the fuel tank.
Fuel Trim	Adjustment made by the PCM as needed to add or remove fuel from its basic fuel map.
HC	Hydrocarbons, unburned fuel from the combustion process
HEGO	Heated Exhaust Gas Oxygen sensor, same as EGO but with an electric heating element.
IAC	Idle Air Control, air passages that bypass the throttle are adjusted by the PCM control idle speed
IAT	Intake Air Temperature sensor, same as ACT, MAT.
KS	Knock Sensor, generates an electrical signal to tell the computer when detonation occurs.

Term	Definition
LED	Light Emitting Diode, solid state device (light) used as an indicator light.
MAF	Mass Airflow sensor, used by the PCM to measure the amount and density of intake air.
MAP	Manifold Absolute Pressure sensor, measures intake manifold vacuum to sense engine load.
MAT	Manifold Air Temperature sensor, same as ACT and IAT.
MIL	Malfunction Indicator Lamp, controlled by the PCM, to warn driver of impending doom.
MPI or MPFI	Multiport Fuel Injection, EFI system with one injector per cylinder.
NO_x	Oxides of Nitrogen, pollutant that occurs when the combustion temperatures are too high.
O_2 Heater	An electric heating element integral to the O_2 sensor.
O_2 Sensor	Measures exhaust gas oxygen content in the engine's exhaust.
PCV	Positive Crankcase Ventilation, directs crankcase gases into the intake manifold.
PCM	Powertrain Control Module, onboard vehicle computer, see ECM.
PFI	Port Fuel Injection, individual fuel injector for each cylinder, see MPI or MPFI.
TBI	Throttle Body Injection, EFI system that uses one or two injectors for the entire engine.
TPS	Throttle Position Sensor, measures throttle opening or angle.
TCC	Torque Converter Clutch, applies a mechanical lock between the engine and auto transmission.
VSS	Vehicle Speed Sensor, measures vehicle road speed.
WOT	Wide Open Throttle, pedal to the metal.

All manufacturers are now required to use this termonology in their service manuals and other technical publications.

Standardized OBD Acronyms

Definition	Acronym
3-2 Timing Solenoid	3-2TS
Four-Wheel Drive Low	4X4L
Accelerator Pedal	AP
Accel. Pedal Position	APP
Air Cleaner	ACl
Air Conditioning	A/C
Air Condo Clutch	ACC
Air Condo Clutch Switch	ACCS
Air Condo Demand	ACD
Air Conditioning ON	ACON
Ambient Air Temperature	AAT
Air Ride Control	ARC
Automatic Transaxle	A/T
Automatic Transmission	A/T
Barometric Pressure	BARO
Battery Positive Voltage	B+
Blower	BlR
Brake ON/OFF	BOO
Brake Pedal Position	BPP
Calculated Load Value	ClV
Camshaft Position	CMP
Canister Purge	CANP
Carburetor	CARB
Central Multipoint Fuel Injector	CMFI
Charge Air Cooler	CAC
Closed Loop	CLL
Closed Throttle Position	CTP
Clutch Pedal Position	CPP
Coast Clutch Solenoid	CCS
Computer-Controlled Dwell	CCD
Constant Control Relay	CCRM
Engine Speed	RPM
Module Erasable Programmable	EPROM
Continuous Fuel Injection	CFI
Read-Only Memory	ROM
Continuous Trap Oxidizer	CTOX
Evaporative Emission	EVAP
Crankshaft Position	CKP
Exhaust Gas Recirculation	EGR
Critical Flow Venturi	CFV
Exhaust Pressure	EP
Cylinder Identification	CID
Fan Control	FC
Data Link Connector	DLC
Flash Electrically Erasable	FEEPROM
Programmable Read Only Memory	PROM
Data Output Line	DOL
Data Negative	DATA-
Flexible Fuel	FF
Data Positive	DATA+
Fourth Gear	4GR
Diagnostic Trouble Code	DTC
Freeze Frame	FRZF
Direct Fuel Injection	DFI
Front-Wheel Drive	FWD
Distributor Ignition	DI
Fuel Level Sensor	FLS
Early Fuel Evaporation	EFE
Fuel Pressure	FP
EGR Temperature	EGRT
Fuel Pump	FP
Electrically Erasable Programmable Read-Only Memory	EEPROM
Fuel Pump Module	FPM
Fuel System Status	Fuel SYS
Electronic Ignition	EI
Fuel Trim	FT
Engine Control	EC
Generator/Alternator	GEN
Engine Control Module	ECM
Grams per Mile	GPM
Engine Coolant Level	ECL
Ground	GND
Engine Coolant Temperature	ECT
Heated Oxygen Sensor	HO_2S
Engine Modification	EM
Idle Air Control	lAC
Engine Oil Pressure	EOP
Idle Speed Control	ISC
Engine Oil Temperature	EOT
Ignition Control	IC
Ignition Control Module	ICM
Park-Neutral Position	PNP
Indirect Fuel Injection	IFI
Parameter Identification	PID
Inertia Fuel Shutoff	IFS
Positive Crankcase Ventilation	PCV
Input Shaft Speed	ISS
Power Steering Pressure	PSP
Inspection and Maintenance	I/M

Definition	Acronym
Intake Air	IA
Powertrain Control Module	PCM
Intake Air Temperature	IAT
Programmable Read-Only Memory	PROM
Intake Manifold Runner Control	IMRC
Pulsed Secondary Air	PAIR
Knock Sensor	KS
Injection Pulse Width Modulated	PWM
Malfunction Indicator Lamp	MIL
Random Access Memory	RAM
Manifold Absolute Pressure	MAP
Read-Only Memory	ROM
Manifold Differential Pressure	MDP
Rear-Wheel Drive	RWD
Manifold Surface	MST
Relay Module	RM
Scan Tool	ST
Manifold Vacuum Zone	MVZ
Secondary Air Injection	AIR
Mass Airflow	MAF
Selectable Four-Wheel Drive	S4WD
Mixture Control	MC
Multipoint Fuel Injection	MFI
Sequential Multipoint	SFI
Nonvolatile Random Access Memory	NVRAM
Fuel Injection	FI
Service Reminder Indication	SRI
Onboard Diagnostics	OBD
Shift Solenoid	SS
Open Loop	OL
Short Term Fuel Trim	STFT
Output Shaft Speed	OSS
Smoke Puff Limiter	SPL
Oxidation Catalytic Converter	OC
Spark Advance	SPKADV
Oxygen Sensor	02S
Supercharger	SC
Supercharger Bypass	SCB
Trans. Fluid Temperature	TFT
System Readiness Test	SRT
Transmission Range	TR
Third Gear	3GR
Turbine Shaft Speed	TSS
Three-Way Catalyst	TWC
Turbocharger	TC
Throttle Body	TB
Vane Airflow	VAF
Throttle Body Fuel Injection	TBI
Variable Control Relay Module	VCRM

Definition	Acronym
Throttle Actuator Control	TAC
Vehicle Control Module	VCM
Throttle Position	TP
Vehicle Identification	VIN
Throttle Position Sensor	TPSensor
Throttle Position Switch	TPSwitch
Vehicle Speed Sensor	VSS
Torque Converter Clutch	TCC
Voltage Regulator	VR
Torque Converter Clutch Pressure	TCCP
Warm-up Oxidation	WU-OC
Catalytic Converter	CAT
Transmission Control	TCM
Warm-Up Three-Way Catalytic Converter	WUTWC
Transmission Fluid Pressure	TFP
Wide Open Throttle	WOT

Just like many drivers don't feel that all traffic regulations and laws apply to them, auto manufacturers take this same view when it comes to the location of their OBD-II diagnostic link connectors, or DLCs. Here is a chart that illustrates hiding places for some of the nonstandard, but allowed, locations for the DLC.

Nonstandard Locations for DLC

Manufacturer	Model(s)	Year(s)	DLC Location
Acura	CL	1997–1998	Under dash, passenger side near center console
Acura	CL	1999	Uncovered, above shifter
Acura	RL	1999–2000	Center console, forward of shifter behind cover
Acura	TL	1996–1998	Center console, behind ashtray
Acura	TL	1999–2000	Behind center dash/console, below stereo, near seat heater control at left
Acura	Integra	1996–1999	Under dash, passenger side near center console
Acura	NSX S2000	1999–2000	Under dash, passenger side near center console
Acura	RL	1996–1998	Front of center console, passenger side
Audi	A4, A4 Avant,	1996	Center console, behind rear sliding ashtray cover
Audi	A6	1996–1997	Center console, behind front tray
Bentley	All	1996–2000	In glove box, behind cover
BMW	3 Series 5 Series, M3	1996–2000	Behind left side of lower left dash but covered by panel. Turn slotted screw 1/4 turn to open
BMW	7 Series	1996–2000	Behind center dash/console, under stereo controls
BMW	X3, M Roadster	1996–2000	Behind passenger side of center dash, console
BMW	Z3	1996–2000	Behind cover, under dash on passenger side
Ferrari	All	1996–2000	Very high under dash, driver's side near center of car
Ford	Bronco	1996	Under dash, slightly right of center
Ford	Trucks, F-Series		Covered
Ford	Thunderbird	1996–1997	Under dash, slightly right of center covered
Honda	Accord	1996–1997	Behind ashtray, center console
Honda	CR-V, Prelude	1997–2000	Under dash, passenger side near center
Honda	Del Sol, Insight	1996–2000	Under dash, passenger side near center
Honda	Odyssey	1996–1998	Behind passenger side of center dash/console
Honda	Prelude	1996	Uncovered, above shifter
Hyundai	Accent	1996–1998	Right center dash, in coin holder
Land Rover	Defender	1997	Left center of dash, under tray
Land Rover	Range Rover	1996–2000	Under right dash, behind cover
Lotus	Espirit	1997–2000	Under cover, above right center of dash/console
Porsche	All	1996	Behind center dash, toward left side
Rolls-Royce	All	1996–2000	In glove box, behind cover
Toyota	Prius	2000	Behind right center dash/console
Toyota	Previa	1996–1997	Behind cover, right side of instrument cluster
Volvo	850	1997–1998	Center console, behind coin holder, forward of shifter
Volvo	All except 850	1998–1999	Behind right side of center console near hand brake
Volvo	S40, V40	2000	Behind cover, left center dash/console
Volvo	C70,S70,V70	2000	Behind cover, center console forward of shifter
Volkswagen	Cabrio	1996–1998	Behind right center dash, to right of ashtray
Volkswagen	Eurovan	1996–early 1999	Under cover, right of instrument cluster, behind wiper control lever
Volkswagen	Golf, Jetta	1999	Behind right center dash
Volkswagen	Passat	1996–1997	Under cover, right of instrument

Diagnostic Trouble Codes

The following tables are a partial list of OBD-II Powertrain Diagnostic Trouble Codes and their definitions. These cover the following systems and/or emission controls:

- **P00XX, Fuel and Air Metering and Auxiliary Emission Controls Codes**
- **P01XX, Fuel and Air Metering Codes**
- **P02XX, Fuel and Air Metering Codes**
- **P03XX, Ignition System or Misfire Codes**
- **P04XX, Auxiliary Emission Control Codes**
- **P05XX, Vehicle Speed, Idle Control, and Auxiliary Input Codes**
- **P06XX, Computer and Auxiliary Output Codes**

P00XX Fuel and Air Metering and Auxiliary Emission Controls DTCs

DTC	Definition
P0001	Fuel Volume Regulator Control Circuit/Open
P0002	Fuel Volume Regulator Control Circuit Range/Performance
P0003	Fuel Volume Regulator Control Circuit Low
P0004	Fuel Volume Regulator Control Circuit High
P0005	Fuel Shutoff Valve "A" Control Circuit/Open
P0006	Fuel Shutoff Valve "A" Control Circuit Low
P0007	Fuel Shutoff Valve "A" Control Circuit High
P0008	Engine Position System Performance, Bank 1
P0009	Engine Position System Performance, Bank 2
P0010*	"A" Camshaft Position Actuator Circuit, Bank 1
P0011*	"A" Camshaft Position-Timing Over-Advanced or System Performance, Bank 1
P0012*	"A" Camshaft Position-Timing Over-Retarded, Bank 1
P0013†	"B" Camshaft Position-Actuator Circuit, Bank 1
P0014†	"B" Camshaft Position-Timing Over-Advanced or System Performance, Bank 1
P0015†	"B" Camshaft Position-Timing Over-Retarded, Bank 1
P0016	Crankshaft Position-Camshaft Position Correlation, Bank 1 Sensor A
P0017	Crankshaft Position-Camshaft Position Correlation, Bank 1 Sensor B
P0018	Crankshaft Position-Camshaft Position Correlation, Bank 2 Sensor A
P0019	Crankshaft Position-Camshaft Position Correlation, Bank 2 Sensor B
P0020*	"A" Camshaft Position Actuator Circuit, Bank 2
P0021*	"A" Camshaft Position-Timing Over-Advanced or System Performance, Bank 2
P0022*	"A" Camshaft Position-Timing Over-Retarded, Bank 2
P0023t	"B" Camshaft Position-Actuator Circuit, Bank 2
P0024t	"B" Camshaft Position-Timing Over-Advanced or System Performance, Bank 2
P0025t	"B" Camshaft Position-Timing Over-Retarded, Bank 2
P0026	Intake Valve Control Solenoid Circuit Range/Performance, Bank 1
P0027	Exhaust Valve Control Solenoid Circuit Range/Performance, Bank 1
P0028	Intake Valve Control Solenoid Circuit Range/Performance, Bank 2
P0029	Exhaust Valve Control Solenoid Circuit Range/Performance, Bank 2
P0030	H02S Heater Control Circuit, Bank 1 Sensor 1
P0031	H02S Heater Control Circuit Low, Bank 1 Sensor 1
P0032	H02S Heater Control Circuit High, Bank 1 Sensor 1
P0033	Turbo Charger Bypass Valve Control Circuit
P0034	Turbo Charger Bypass Valve Control Circuit Low
P0035	Turbo Charger Bypass Valve Control Circuit High

DTC	Definition
P0036	HO_2S Heater Control Circuit, Bank 1 Sensor 2
P0037	HO_2S Heater Control Circuit Low, Bank 1 Sensor 2
P0038	HO_2S Heater Control Circuit High, Bank 1 Sensor 2
P0039	Turbo/Super Charger Bypass Valve Control Circuit Range/Performance
P0040	O_2 Sensor Signals Swapped, Bank 1 Sensor 1 /Bank 2 Sensor 1
P0041	O_2 Sensor Signals Swapped, Bank 1 Sensor 2/Bank 2 Sensor 2
P0042	HO_2S Heater Control Circuit, Bank 1 Sensor 3
P0043	HO_2S Heater Control Circuit Low, Bank 1 Sensor 3
P0044	HO_2S Heater Control Circuit High, Bank 1 Sensor 3
P0045	Turbo/Super Charger Boost Control Solenoid Circuit/Open
P0046	Turbo/Super Charger Boost Control Solenoid Circuit Range/Performance
P0047	Turbo/Super Charger Boost Control Solenoid Circuit Low
P0048	Turbo/Super Charger Boost Control Solenoid Circuit High
P0049	Turbo/Super Charger Turbine Overspeed
P0050	HO_2S Heater Control Circuit, Bank 2 Sensor 1
P0051	HO_2S Heater Control Circuit Low, Bank 2 Sensor 1
P0052	HO_2S Heater Control Circuit High, Bank 2 Sensor 1
P0053	HO_2S Heater Resistance, Bank 1 Sensor 1
P0054	HO_2S Heater Resistance, Bank 1 Sensor 2
P0055	HO_2S Heater Resistance, Bank 1 Sensor 3
P0056	HO_2S Heater Control Circuit, Bank 2 Sensor 2
P0057	HO_2S Heater Control Circuit Low, Bank 2 Sensor 2
P0058	HO_2S Heater Control Circuit High, Bank 2 Sensor 2
P0059	HO_2S Heater Resistance, Bank 2 Sensor 1
P0060	HO_2S Heater Resistance, Bank 2 Sensor 2
P0061	HO_2S Heater Resistance, Bank 2 Sensor 3
P0062	HO_2S Heater Control Circuit, Bank 2 Sensor 3
P0063	HO_2S Heater Control Circuit Low, Bank 2 Sensor 3
P0064	HO_2S Heater Control Circuit High, Bank 2 Sensor 3
P0065	Air Assisted Injector Control Range/Performance
P0066	Air Assisted Injector Control Circuit or Circuit Low
P0067	Air Assisted Injector Control Circuit High
P0068	MAP /MAF Throttle Position Correlation
P0069	Manifold Absolute Pressure-Barometric Pressure Correlation
P0070	Ambient Air Temperature Sensor Circuit
P0071	Ambient Air Temperature Sensor Range/Performance
P0072	Ambient Air Temperature Sensor Circuit Low
P0073	Ambient Air Temperature Sensor Circuit High
P0074	Ambient Air Temperature Sensor Circuit Intermittent
P0075	Intake Valve Control Solenoid Circuit, Bank 1
P0076	Intake Valve Control Solenoid Circuit Low, Bank 1
P0077	Intake Valve Control Solenoid Circuit High, Bank 1
P0078	Exhaust Valve Control Solenoid Circuit, Bank 1
P0079	Exhaust Valve Control Solenoid Circuit Low, Bank 1
P0080	Exhaust Valve Control Solenoid Circuit high, Bank 1
P0081	Intake Valve Control Solenoid Circuit, Bank 2
P0082	Intake Valve Control Solenoid Circuit Low, Bank 2
P0083	Intake Valve Control Solenoid Circuit High, Bank 2
P0084	Exhaust Valve Control Solenoid Circuit, Bank 2

DTC	Definition
P0085	Exhaust Valve Control Solenoid Circuit Low, Bank 2
P0086	Exhaust Valve Control Solenoid Circuit High, Bank 2
P0087	Fuel Rail/System Pressure—Too Low
P0088	Fuel Rail/System Pressure—Too High
P0089	Fuel Pressure Regulator 1 Performance
P0090	Fuel Pressure Regulator 1 Control Circuit
P0091	Fuel Pressure Regulator 1 Control Circuit Low
P0092	Fuel Pressure Regulator 1 Control Circuit High
P0093	Fuel System Leak Detected—Large Leak
P0094	Fuel System Leak Detected—Small Leak
P0095	Intake Air Temperature Sensor 2 Circuit
P0096	Intake Air Temperature Sensor 2 Circuit Range/Performance
P0097	Intake Air Temperature Sensor 2 Circuit Low
P0098	Intake Air Temperature Sensor 2 Circuit High
P0099	Intake Air Temperature Sensor 2 Circuit Intermittent/Erratic

**The "A" camshaft can be either the intake, left, or front camshaft. Left/right and front/rear are determined as if viewed from the driver's seat. Bank 1 contains cylinder number one; Bank 2 is the opposite bank on the engine.*

†The "B" camshaft can be either the exhaust, right, or rear camshaft. Left/right and front/rear are determined as if viewed from the driver's seat. Bank 1 contains cylinder number one; Bank 2 is the opposite bank of the engine.

P01XX Fuel and Air Metering DTCs

DTC	Definition
P0100	Mass or Volume Airflow Circuit
P0101	Mass or Volume Airflow Circuit Range/Performance
P0102	Mass or Volume Airflow Circuit Low Input
P0103	Mass or Volume Airflow Circuit High Input
P0104	Mass or Volume Airflow Circuit Intermittent
P0105	Manifold Absolute Pressure/Barometric Pressure Circuit
P0106	Manifold Absolute Pressure/Barometric Pressure Circuit Range/Performance
P0107	Manifold Absolute Pressure/Barometric Pressure Circuit Low Input
P0108	Manifold Absolute Pressure/Barometric Pressure Circuit High Input
P0109	Manifold Absolute Pressure/Barometric Pressure Circuit Intermittent
PO110	Intake Air Temperature Sensor 1 Circuit
PO111	Intake Air Temperature Sensor 1 Circuit Range/Performance
PO112	Intake Air Temperature Sensor 1 Circuit Low
PO113	Intake Air Temperature Sensor 1 Circuit High
PO114	Intake Air Temperature Sensor 1 Circuit Intermittent
PO115	Engine Coolant Temperature Circuit
PO116	Engine Coolant Temperature Circuit Range/Performance
PO117	Engine Coolant Temperature Circuit Low
PO118	Engine Coolant Temperature Circuit High
PO119	Engine Coolant Temperature Circuit Intermittent
PO120	Throttle/Pedal Position Sensor/Switch "A" Circuit
PO121	Throttle/Pedal Position Sensor/Switch "A" Circuit Range/Performance
PO122	Throttle/Pedal Position Sensor/Switch "A" Circuit Low

DTC	Definition
PO123	Throttle/Pedal Position Sensor/Switch "A" Circuit High
PO124	Throttle/Pedal Position Sensor/Switch "A" Circuit Intermittent
PO125	Insufficient Coolant Temperature for Closed Loop Fuel Control
PO126	Insufficient Coolant Temperature for Stable Operation
PO127	Intake Air Temperature Too High
PO128	Coolant Thermostat (Coolant Temperature below Thermostat Regulating Temperature)
PO129	Barometric Pressure Too Low
PO130	O_2 Sensor Circuit, Bank 1 Sensor 1
PO131	O_2 Sensor Circuit Low Voltage, Bank 1 Sensor 1
PO132	O_2 Sensor Circuit High Voltage, Bank 1 Sensor 1
PO133	O_2 Sensor Circuit High Voltage, Bank 1 Sensor 1
PO134	O_2 Sensor Circuit No Activity Detected, Bank 1 Sensor 1
PO135	O_2 Sensor Heater Circuit, Bank 1 Sensor 1
PO136	O_2 Sensor Circuit, Bank 1 Sensor 2
PO137	O_2 Sensor Circuit Low Voltage, Bank 1 Sensor 2
PO138	O_2 Sensor Circuit High Voltage, Bank 1 Sensor 2
PO139	O_2 Sensor Circuit Slow Response, Bank 1 Sensor 2
PO140	O_2 Sensor Circuit No Activity Detected, Bank 1 Sensor 2
PO141	O_2 Sensor Heater Circuit, Bank 1 Sensor 2
PO142	O_2 Sensor Circuit, Bank 1 Sensor 3
PO143	O_2 Sensor Circuit Low Voltage, Bank 1 Sensor 3
PO144	O_2 Sensor Circuit High Voltage, Bank 1 Sensor 3
PO145	O_2 Sensor Circuit Slow Response, Bank 1 Sensor 3
PO146	O_2 Sensor Circuit No Activity Detected, Bank 1 Sensor 3
PO147	O_2 Sensor Heater Circuit, Bank 1, Sensor 3
PO148	Fuel Delivery Error
PO149	Fuel Timing Error
PO150	O_2 Sensor Circuit, Bank 2 Sensor 1
PO151	O_2 Sensor Circuit Low Voltage, Bank 2 Sensor 1
PO152	O_2 Sensor Circuit High Voltage, Bank 2 Sensor 1
PO153	O_2 Sensor Circuit Slow Response, Bank 2 Sensor 1
PO154	O_2 Sensor Circuit No Activity Detected, Bank 2 Sensor 1
PO155	O_2 Sensor Heater Circuit, Bank 2 Sensor 1
PO156	O_2 Sensor Circuit, Bank 2 Sensor 2
PO157	O_2 Sensor Circuit Low Voltage, Bank 2 Sensor 2
PO158	O_2 Sensor Circuit High Voltage, Bank 2 Sensor 2
PO159	O_2 Sensor Circuit Slow Response, Bank 2 Sensor 2
PO160	O_2 Sensor Circuit No Activity Detected, Bank 2 Sensor 2
P0161	O_2 Sensor Heater Circuit, Bank 2 Sensor 2
P0162	O_2 Sensor Circuit, Bank 2 Sensor 3
P0163	O_2 Sensor Circuit Low Voltage, Bank 2 Sensor 3
P0164	O_2 Sensor Circuit High Voltage, Bank 2 Sensor 3
P0165	O_2 Sensor Circuit Slow Response, Bank 2 Sensor 3
P0166	O_2 Sensor Circuit No Activity Detected, Bank 2 Sensor 3
P0167	O_2 Sensor Heater Circuit, Bank 2 Sensor 3
P0168	Fuel Temperature Too High
P0169	Incorrect Fuel Consumption
P0170	Fuel Trim, Bank 1
P0171	System Too Lean, Bank 1

DTC	Definition
P0172	System Too Rich, Bank 1
P0173	Fuel Trim, Bank 2
P0174	System Too Lean, Bank 2
P017S	System Too Rich, Bank 2
P0176	Fuel Composition Sensor Circuit
P0177	Fuel Composition Sensor Circuit Range/Performance
P0178	Fuel Composition Sensor Circuit Low
P0179	Fuel Composition Sensor Circuit High
P0180	Fuel Temperature Sensor A Circuit
P0181	Fuel Temperature Sensor A Circuit Range/Performance
P0182	Fuel Temperature Sensor A Circuit Low
P0183	Fuel Temperature Sensor A Circuit High
P0184	Fuel Temperature Sensor A Circuit Intermittent
P0185	Fuel Temperature Sensor B Circuit
P0186	Fuel Temperature Sensor B Circuit Range/Performance
P0187	Fuel Temperature Sensor B Circuit Low
P0188	Fuel Temperature Sensor B Circuit High
P0189	Fuel Temperature Sensor B Circuit Intermittent
P0190	Fuel Rail Pressure Sensor Circuit
P0191	Fuel Rail Pressure Sensor Circuit Range/Performance
P0192	Fuel Rail Pressure Sensor Circuit Low
PO193	Fuel Rail Pressure Sensor Circuit High
PO194	Fuel Rail Pressure Sensor Circuit Intermittent
PO195	Engine Oil Temperature Sensor
PO196	Engine Oil Temperature Sensor Range/Performance
PO197	Engine Oil Temperature Sensor Low
PO198	Engine Oil Temperature Sensor High
PO199	Engine Oil Temperature Sensor Intermittent

P02XX Fuel and Air Metering DTCs

DTC	Definition
P0200	Injector Circuit/Open
P0201	Injector Circuit/Open—Cylinder 1
P0202	Injector Circuit/Open—Cylinder 2
P0203	Injector Circuit/Open-—Cylinder 3
P0204	Injector Circuit/Open—Cylinder 4
P0205	Injector Circuit/Open—Cylinder 5
P0206	Injector Circuit/Open—Cylinder 6
P0207	Injector Circuit/Open—Cylinder 7
P0208	Injector Circuit/Open—Cylinder 8
P0209	Injector Circuit/Open—Cylinder 9
P0210	Injector Circuit/Open—Cylinder 10
P0211	Injector Circuit/Open—Cylinder 11
P0212	Injector Circuit/Open—Cylinder 12
P0213	Cold Start Injector 1
P0214	Cold Start Injector 2
P0215	Engine Shutoff Solenoid

DTC	Definition
P0216	Injector/Injection Timing Control Circuit
P0217	Engine Coolant over Temperature Condition
P0218	Transmission Fluid over Temperature Condition
P0219	Engine Overspeed Condition
P0220	Throttle/Pedal Position Sensor/Switch "B" Circuit
P0221	Throttle/Pedal Position Sensor/Switch "B" Circuit Range/Performance
P0222	Throttle/Pedal Position Sensor/Switch "B" Circuit Low
P0223	Throttle/Pedal Position Sensor/Switch "B" Circuit High
P0224	Throttle/Pedal Position Sensor/Switch "B" Circuit Intermittent
P0225	Throttle/Pedal Position Sensor/Switch "C" Circuit
P0226	Throttle/Pedal Position Sensor/Switch "C" Circuit Range/Performance
P0227	Throttle/Pedal Position Sensor/Switch "E" Circuit Low
P0228	Throttle/Pedal Position Sensor/Switch "E" Circuit High
P0229	Throttle/Pedal Position Sensor/Switch "E" Circuit Intermittent
P0230	Fuel Pump Primary Circuit
P0231	Fuel Pump Primary Circuit Low
P0232	Fuel Pump Primary Circuit High
P0233	Fuel Pump Primary Circuit Intermittent
P0234	Turbo/Super Charger Overboost Condition
P0235	Turbo/Super Charger Boost Sensor "A" Circuit
P0236	Turbo/Super Charger Boost Sensor "A" Circuit Range/Performance
P0237	Turbo/Super Charger Boost Sensor "A" Circuit Low
P0238	Turbo/Super Charger Boost Sensor "A" Circuit High
P0239	Turbo/Super Charger Boost Sensor "B" Circuit
P0240	Turbo/Super Charger Boost Sensor "B" Circuit Range/Performance
P0241	Turbo/Super Charger Boost Sensor "B" Circuit Low
P0242	Turbo/Super Charger Boost Sensor "B" Circuit High
P0243	Turbo/Super Charger Wastegate Solenoid "A"
P0244	Turbo/Super Charger Wastegate Solenoid "A" Range/Performance
P0245	Turbo/Super Charger Wastegate Solenoid "A" Low
P0246	Turbo/Super Charger Wastegate Solenoid "A" High
P0247	Turbo/Super Charger Wastegate Solenoid "B"
P0248	Turbo/Super Charger Wastegate Solenoid "B" Range/Performance
P0249	Turbo/Super Charger Wastegate Solenoid "B" Low
P0250	Turbo/Super Charger Wastegate Solenoid "B" High
P0251	Injection Pump Fuel Metering Control "A" (Cam/Rotor/Injector)
P0252	Injection Pump Fuel Metering Control "A" Range/Performance (Cam/Rotor/Injector)
P0253	Injection Pump Fuel Metering Control "A" Low (Cam/Rotor/Injector)
P0254	Injection Pump Fuel Metering Control "A" High (Cam/Rotor/Injector)
P0255	Injection Pump Fuel Metering Control "A" Intermittent (Cam/Rotor/Injector)
P0256	Injection Pump Fuel Metering Control "B" (Cam/Rotor/Injector)
P0257	Injection Pump Fuel Metering Control "B" Range/Performance (Cam/Rotor/Injector)
P0258	Injection Pump Fuel Metering Control "B" Low (Cam/Rotor/Injector)
P0259	Injection Pump Fuel Metering Control "B" High (Cam/Rotor/Injector)
P0260	Injection Pump Fuel Metering Control "B" Intermittent (Cam/Rotor/Injector)
P0261	Cylinder 1 Injector Circuit Low
P0262	Cylinder 1 Injector Circuit High
P0263	Cylinder 1 Contribution/Balance
P0264	Cylinder 2 Injector Circuit Low

DTC	Definition
P0265	Cylinder 2 Injector Circuit High
P0266	Cylinder 2 Contribution/Balance
P0267	Cylinder 3 Injector Circuit Low
P0268	Cylinder 3 Injector Circuit High
P0269	Cylinder 3 Contribution/Balance
P0270	Cylinder 4 Injector Circuit Low
P0271	Cylinder 4 Injector Circuit High
P0272	Cylinder 4 Contribution/Balance
P0273	Cylinder 5 Injector Circuit Low
P0274	Cylinder 5 Injector Circuit High
P0275	Cylinder 5 Contribution/Balance
P0276	Cylinder 6 Injector Circuit Low
P0277	Cylinder 6 Injector Circuit High
P0278	Cylinder 6 Contribution/Balance
P0279	Cylinder 7 Injector Circuit Low
P0280	Cylinder 7 Injector Circuit High
P0281	Cylinder 7 Contribution/Balance
P0282	Cylinder 8 Injector Circuit Low
P0283	Cylinder 8 Injector Circuit High
P0284	Cylinder 8 Contribution/Balance
P0285	Cylinder 9 Injector Circuit Low
P0286	Cylinder 9 Injector Circuit High
P0287	Cylinder 9 Contribution/Balance
P0288	Cylinder 10 Injector Circuit Low
P0289	Cylinder 10 Injector Circuit High
P0290	Cylinder 10 Contribution/Balance
P0291	Cylinder 11 Injector Circuit Low
P0292	Cylinder 11 Injector Circuit High
P0293	Cylinder 11 Contribution/Balance
P0294	Cylinder 12 Injector Circuit Low
P0295	Cylinder 12 Injector Circuit High
P0296	Cylinder 12 Contribution/Balance
P0297	Vehicle Overspeed Condition
P0298	Engine Oil over Temperature
P0299	Turbo/Super Charger Underboost

P03XX Ignition System or Misfire DTCs

DTC	Definition
P0300	Random/Multiple Cylinder Misfire Detected
P0301	Cylinder 1 Misfire Detected
P0302	Cylinder 2 Misfire Detected
P0303	Cylinder 3 Misfire Detected
P0304	Cylinder 4 Misfire Detected
P0305	Cylinder 5 Misfire Detected
P0306	Cylinder 6 Misfire Detected
P0307	Cylinder 7 Misfire Detected
P0308	Cylinder 8 Misfire Detected

DTC	Definition
P0309	Cylinder 9 Misfire Detected
P0310	Cylinder 10 Misfire Detected
P0311	Cylinder 11 Misfire Detected
P0312	Cylinder 12 Misfire Detected
P0313	Misfire Detected with Low Fuel
P0314	Single Cylinder Misfire (Cylinder Not Specified)
P0315	Crankshaft Position System Variation Not Learned
P0316	Engine Misfire Detected on Startup (First 1,000 Revolutions)
P0317	Rough Road Hardware Not Present
P0318	Rough Road Sensor "A" Signal Circuit
P0319	Rough Road Sensor "B"
P0320	Ignition/Distributor Engine Speed Input Circuit
P0321	Ignition/Distributor Engine Speed Input Circuit Range/Performance
P0322	Ignition/Distributor Engine Speed Input Circuit No Signal
P0323	Ignition/Distributor Engine Speed Input Circuit Intermittent
P0324	Knock Control System Error
P0325	Knock Sensor 1 Circuit, Bank 1 or Single Sensor
P0326	Knock Sensor 1 Circuit Range/Performance, Bank 1 or Single Sensor
P0327	Knock Sensor 1 Circuit Low, Bank 1 or Single Sensor
P0328	Knock Sensor 1 Circuit High, Bank 1 or Single Sensor
P0329	Knock Sensor 1 Circuit Input Intermittent, Bank 1 or Single Sensor
P0330	Knock Sensor 2 Circuit, Bank 2
P0331	Knock Sensor 2 Circuit Range/Performance, Bank 2
P0332	Knock Sensor 2 Circuit Low, Bank 2
P0333	Knock Sensor 2 Circuit High, Bank 2
P0334	Knock Sensor 2 Circuit Input Intermittent, Bank 2
P0335	Crankshaft Position Sensor "A" Circuit
P0336	Crankshaft Position Sensor "A" Circuit Range/Performance
P0337	Crankshaft Position Sensor "A" Circuit Low
P0338	Crankshaft Position Sensor "A" Circuit High
P0339	Crankshaft Position Sensor "A" Circuit Intermittent
P0340	Crankshaft Position Sensor "A" Circuit, Bank 1 or Single Sensor
P0341	Crankshaft Position Sensor" A" Circuit Range/Performance, Bank 1 or Single Sensor
P0342	Crankshaft Position Sensor "A" Circuit Low, Bank 1 or Single Sensor
P0343	Crankshaft Position Sensor "A" Circuit High, Bank 1 or Single Sensor
P0344	Crankshaft Position Sensor II "A" Circuit Intermittent, Bank 1 or Single Sensor
P0345	Crankshaft Position Sensor "A" Circuit, Bank 2
P0346	Crankshaft Position Sensor "A" Circuit Range/Performance, Bank 2
P0347	Crankshaft Position Sensor "A" Circuit Low, Bank 2
P0348	Crankshaft Position Sensor "A" Circuit High, Bank 2
P0349	Crankshaft Position Sensor "A" Circuit Intermittent, Bank 2
P0350	Ignition Coil Primary/Secondary Circuit
P0351	Ignition Coil "A" Primary/Secondary Circuit
P0352	Ignition Coil "B" Primary/Secondary Circuit
P0353	Ignition Coil "C" Primary/Secondary Circuit
P0354	Ignition Coil "D" Primary/Secondary Circuit
P0355	Ignition Coil "E" Primary/Secondary Circuit
P0356	Ignition Coil "F" Primary/Secondary Circuit
P0357	Ignition Coil "G" Primary/Secondary Circuit

DTC	Definition
P0358	Ignition Coil "H" Primary/Secondary Circuit
P0359	Ignition Coil "I" Primary/Secondary Circuit
P0360	Ignition Coil "J" Primary/Secondary Circuit
P0361	Ignition Coil "K" Primary/Secondary Circuit
P0362	Ignition Coil "L" Primary/Secondary Circuit
P0363	Misfire Detected-Fueling Disabled
P0364	Reserved
P0365	Camshaft Position Sensor "B" Circuit, Bank 1
P0366	Camshaft Position Sensor "B" Circuit Range/Performance, Bank 1
P0367	Camshaft Position Sensor "B" Circuit Low, Bank 1
P0368	Camshaft Position Sensor "B" Circuit High, Bank 1
P0369	Camshaft Position Sensor "B" Circuit Intermittent, Bank 1
P0370	Timing Reference High Resolution Signal "A"
P0371	Timing Reference High Resolution Signal "A" Too Many Pulses
P0372	Timing Reference High Resolution Signal "A" Too Few Pulses
P0373	Timing Reference High Resolution Signal "A" Intermittent/Erratic Pulses
P0374	Timing Reference High Resolution Signal "A" No Pulse
P0375	Timing Reference High Resolution Signal "B"
P0376	Timing Reference High Resolution Signal "B" Too Many Pulses
P0377	Timing Reference High Resolution Signal "B" Too Few Pulses
P0378	Timing Reference High Resolution Signal "B" Intermittent/Erratic Pulses
P0379	Timing Reference High Resolution Signal "B" No Pulses
P0380	Glow Plug/Heater Circuit "A"
P0381	Glow Plug/Heater Indicator Circuit
P0382	Glow Plug/Heater Circuit "B"
P0383	Reserved by document
P0384	Reserved by document
P0385	Crankshaft Position Sensor "B" Circuit
P0386	Crankshaft Position Sensor "B" Circuit Range/Performance
P0387	Crankshaft Position Sensor "B" Circuit Low
P0388	Crankshaft Position Sensor "B" Circuit High
P0389	Crankshaft Position Sensor "B" Circuit Intermittent
P0390	Crankshaft Position Sensor "B" Circuit, Bank 2
P0391	Crankshaft Position Sensor "B" Circuit Range/Performance, Bank 2
P0392	Crankshaft Position Sensor "B" Circuit Low, Bank 2
P0393	Crankshaft Position Sensor "B" Circuit High, Bank 2
P0394	Crankshaft Position Sensor "B" Circuit Intermittent, Bank 2

P04XX Auxiliary Emission Controls DTCs

DTC	Definition
P0400	Exhaust Gas Recirculation Flow
P0401	Exhaust Gas Recirculation Flow Insufficient Detected
P0402	Exhaust Gas Recirculation Flow Excessive Detected
P0403	Exhaust Gas Recirculation Control Circuit
P0404	Exhaust Gas Recirculation Control Circuit Range/Performance
P0405	Exhaust Gas Recirculation Sensor "A" Circuit Low
P0406	Exhaust Gas Recirculation Sensor "A" Circuit High

DTC	Definition
P0407	Exhaust Gas Recirculation Sensor "B" Circuit Low
P0408	Exhaust Gas Recirculation Sensor "B" Circuit High
P0409	Exhaust Gas Recirculation Sensor "A"
P0410	Secondary Air Injection System
P0411	Secondary Air Injection System Incorrect Flow Detected
P0412	Secondary Air Injection System Switching Valve "A" Circuit
P0413	Secondary Air Injection System Switching Valve "A" Circuit Open
P0414	Secondary Air Injection System Switching Valve" A" Circuit Shorted
P0415	Secondary Air Injection System Switching Valve "B" Circuit
P0416	Secondary Air Injection System Switching Valve "B" Circuit Open
P0417	Secondary Air Injection System Switching Valve "B" Circuit Shorted
P0418	Secondary Air Injection System Control "A" Circuit
P0419	Secondary Air Injection System Control "B" Circuit
P0420	Catalyst System Efficiency below Threshold, Bank 1
P0421	Warm-Up Catalyst Efficiency below Threshold, Bank 1
P0422	Main Catalyst Efficiency below Threshold, Bank 1
P0423	Heated Catalyst Efficiency below Threshold, Bank 1
P0424	Heated Catalyst Temperature below Threshold, Bank 1
P0425	Catalyst Temperature Sensor, Bank 1
P0426	Catalyst Temperature Sensor Range/Performance, Bank 1
P0427	Catalyst Temperature Sensor Low, Bank 1
P0428	Catalyst Temperature Sensor High, Bank 1
P0429	Catalyst Heater Control Circuit, Bank 1
P0430	Catalyst System Efficiency below Threshold, Bank 2
P0431	Warm Up Catalyst Efficiency below Threshold, Bank 2
P0432	Main Catalyst Efficiency below Threshold, Bank 2
P0433	Heated Catalyst Efficiency below Threshold, Bank 2
P0434	Heated Catalyst Temperature below Threshold, Bank 2
P0435	Catalyst Temperature Sensor, Bank 2
P0436	Catalyst Temperature Sensor Range/Performance, Bank 2
P0437	Catalyst Temperature Sensor Low, Bank 2
P0438	Catalyst Temperature Sensor High, Bank 2
P0439	Catalyst Heater Control Circuit, Bank 2
P0440	Evaporative Emission System
P0441	Evaporative Emission System Incorrect Purge Flow
P0442	Evaporative Emission System Leak Detected (Small Leak)
P0443	Evaporative Emission System Purge Control Valve Circuit
P0444	Evaporative Emission System Purge Control Valve Circuit Open
P0445	Evaporative Emission System Purge Control Valve Circuit Shorted
P0446	Evaporative Emission System Vent Control Circuit
P0447	Evaporative Emission System Vent Control Circuit Open
P0448	Evaporative Emission System Vent Control Circuit Shorted
P0449	Evaporative Emission System Vent Valve/Solenoid Circuit
P0450	Evaporative Emission System Pressure Sensor/Switch
P0451	Evaporative Emission System Pressure Sensor/Switch Range/Performance
P0452	Evaporative Emission System Pressure Sensor/Switch Low
P0453	Evaporative Emission System Pressure Sensor/Switch High
P0454	Evaporative Emission System Pressure Sensor/Switch Intermittent
P0455	Evaporative Emission System Leak Detected (Large Leak)

DTC	Definition
P0456	Evaporative Emission System Leak Detected (Very Small Leak)
P0457	Evaporative Emission System Leak Detected (Fuel Cap Loose/Off)
P0458	Evaporative Emission System Purge Control Valve Circuit Low
P0459	Evaporative Emission System Purge Control Valve Circuit High
P0460	Fuel Level Sensor "A" Circuit
P0461	Fuel Level Sensor "A" Circuit Range/Performance
P0462	Fuel Level Sensor "A" Circuit Low
P0463	Fuel Level Sensor "A" Circuit High
P0464	Fuel Level Sensor "A" Circuit Intermittent
P0465	EVAP Purge Flow Sensor Circuit
P0466	EVAP Purge Flow Sensor Circuit Range/Performance
P0467	EVAP Purge Flow Sensor Circuit Low
P0468	EVAP Purge Flow Sensor Circuit High
P0469	EVAP Purge Flow Sensor Circuit Intermittent
P0470	Exhaust Pressure Sensor
P0471	Exhaust Pressure Sensor Range/Performance
P0472	Exhaust Pressure Sensor Low
P0473	Exhaust Pressure Sensor High
P0474	Exhaust Pressure Sensor Intermittent
P0475	Exhaust Pressure Control Valve
P0476	Exhaust Pressure Control Valve Range/Performance
P0477	Exhaust Pressure Control Valve Low
P0478	Exhaust Pressure Control Valve High
P0479	Exhaust Pressure Control Valve Intermittent
P0480	Fan 1 Control Circuit
P0481	Fan 2 Control Circuit
P0482	Fan 3 Control Circuit
P0483	Fan Rationality Check
P0484	Fan Circuit over Current
P0485	Fan Power/Ground Circuit
P0486	Exhaust Gas Recirculation Sensor "B" Circuit
P0487	Exhaust Gas Recirculation Throttle Position Control Circuit
P0488	Exhaust Gas Recirculation Throttle Position Control Range/Performance
P0489	Exhaust Gas Recirculation Control Circuit Low
P0490	Exhaust Gas Recirculation Control Circuit High
P0491	Secondary Air Injection System Insufficient Flow, Bank 1
P0492	Secondary Air Injection System Insufficient Flow, Bank 2
P0493	Fan Overspeed
P0494	Fan Speed Low
P0495	Fan Speed High
P0496	Evaporative Emission System High Purge Flow
P0497	Evaporative Emission System Low Purge Flow
P0498	Evaporative Emission System Vent Valve Control Circuit Low
P0499	Evaporative Emission System Vent Valve Control Circuit High

P05XX Vehicle Speed, Idle Control, and Auxiliary Input DTCs

DTC	Definition
P0500	Vehicle Speed Sensor "A"
P0501	Vehicle Speed Sensor "A" Range/Performance
P0502	Vehicle Speed Sensor "A" Circuit Low Input
P0503	Vehicle Speed Sensor "A" Intermittent/Erratic/High
P0504	Brake Switch "A"/"B" Correlation
P0505	Idle Air Control System
P0506	Idle Air Control System RPM Lower than Expected
P0507	Idle Air Control System RPM Higher than Expected
P0508	Idle Air Control System Circuit Low
P0509	Idle Air Control System Circuit High
P0510	Closed Throttle Position Switch
P0511	Idle Air Control Circuit
P0512	Starter Request Circuit
P0513	Incorrect Immobilizer Key
P0514	Battery Temperature Sensor Circuit Range/Performance
P0515	Battery Temperature Sensor Circuit
P0516	Battery Temperature Sensor Circuit Low
P0517	Battery Temperature Sensor Circuit High
P0518	Idle Air Control Circuit Intermittent
P0519	Idle Air Control System Performance
P0520	Engine Oil Pressure Sensor/Switch Circuit
P0521	Engine Oil Pressure Sensor/Switch Range/Performance
P0522	Engine Oil Pressure Sensor/Switch Low Voltage
P0523	Engine Oil Pressure Sensor/Switch High Voltage
P0524	Engine Oil Pressure Too Low
P0525	Cruise Control Servo Control Circuit Range/Performance
P0526	Fan Speed Sensor Circuit
P0527	Fan Speed Sensor Circuit Range/Performance
P0528	Fan Speed Sensor Circuit No Signal
P0529	Fan Speed Sensor Circuit Intermittent
P0530	A/C Refrigerant Pressure Sensor "A" Circuit
P0531	A/C Refrigerant Pressure Sensor "A" Circuit Range/Performance
P0532	A/C Refrigerant Pressure Sensor "A" Circuit Low
P0533	A/C Refrigerant Pressure Sensor "A" Circuit High
P0534	Air Conditioner Refrigerant Charge Loss
P0535	A/C Evaporator Temperature Sensor Circuit
P0536	A/C Evaporator Temperature Sensor Circuit Range/Performance
P0537	A/C Evaporator Temperature Sensor Circuit Low
P0538	A/C Evaporator Temperature Sensor Circuit High
P0539	A/C Evaporator Temperature Sensor Circuit Intermittent
P0540*	Intake Air Heater "A" Circuit
P0541*	Intake Air Heater "A" Circuit Low
P0542*	Intake Air Heater "A" Circuit High
P0543*	Intake Air Heater "A" Circuit Open
P0544	Exhaust Gas Temperature Sensor Circuit, Bank 1 Sensor 1
P0545	Exhaust Gas Temperature Sensor Circuit Low, Bank 1 Sensor 1
P0546	Exhaust Gas Temperature Sensor Circuit High, Bank 1 Sensor 1

DTC	Definition
P0547	Exhaust Gas Temperature Sensor Circuit, Bank 2 Sensor 1
P0548	Exhaust Gas Temperature Sensor Circuit Low, Bank 2 Sensor 1
P0549	Exhaust Gas Temperature Sensor Circuit High, Bank 2 Sensor 1
P0550	Power Steering Pressure Sensor/Switch Circuit
P0551	Power Steering Pressure Sensor/Switch Circuit Range/Performance
P0552	Power Steering Pressure Sensor/Switch Circuit Low Input
P0553	Power Steering Pressure Sensor/Switch Circuit High Input
P0554	Power Steering Pressure Sensor/Switch Circuit Intermittent
P0555	Brake Booster Pressure Sensor Circuit
P0556	Brake Booster Pressure Sensor Circuit Range/Performance
P0557	Brake Booster Pressure Sensor Circuit Low Input
P0558	Brake Booster Pressure Sensor Circuit High Input
P0559	Brake Booster Pressure Sensor Circuit Intermittent
P0560	System Voltage
P0561	System Voltage Unstable
P0562	System Voltage Low
P0563	System Voltage High
P0564	Cruise Control Multifunction Input "A" Circuit
P0565	Cruise Control On Signal
P0566	Cruise Control Off Signal
P0567	Cruise Control Resume Signal
P0568	Cruise Control Set Signal
P0569	Cruise Control Coast Signal
P0570	Cruise Control Accelerate Signal
P0571	Brake Switch "A" Circuit
P0572	Brake Switch "A" Circuit Low
P0573	Brake Switch "A" Circuit High
P0574	Cruise Control System—Vehicle Speed Too High
P0575	Cruise Control Input Circuit
P0576	Cruise Control Input Circuit Low
P0577	Cruise Contrallnput Circuit High
P0578†	Cruise Control Multifunction Input "A" Circuit Stuck
P0579†	Cruise Control Multifunction Input "A" Circuit Range/Performance
P0580†	Cruise Control Multifunction Input "A" Circuit Low
P0581†	Cruise Control Multifunction Input "A" Circuit High
P0582	Cruise Control Vacuum Control Circuit/Open
P0583	Cruise Control Vacuum Control Circuit Low
P0584	Cruise Control Vacuum Control Circuit High
P0585	Cruise Control Multifunction Input "A"/"B" Correlation
P0586	Cruise Control Vent Control Circuit/Open
P0587	Cruise Control Vent Control Circuit Low
P0588	Cruise Control Vent Control Circuit High
P0589	Cruise Control Multifunction Input "B" Circuit
P0590	Cruise Control Multifunction Input "B" Circuit Stuck
P0591	Cruise Control Multifunction Input "B" Circuit Range/Performance
P0592	Cruise Control Multifunction Input "B" Circuit Low
P0593	Cruise Control Multifunction Input "B" Circuit High
P0594	Cruise Control Servo Control Circuit/Open
P0595	Cruise Control Servo Control Circuit Low

DTC	Definition
P0596	Cruise Control Servo Control Circuit High
P0597	Thermostat Heater Control Circuit/Open
P0598	Thermostat Heater Control Circuit Low
P0599	Thermostat Heater Control Circuit High

**For DTCs PO540–PO543 also see P2604–P2609*

†For DTCs PO578–PO581 also see PO564

PO6XX Computer and Auxiliary Output DTCs

DTC	Definition
P0600	Serial Communication Link
P0601	Internal Control Module Memory Check Sum Error
P0602	Control Module Programming Error
P0603	Internal Control Module Keep Alive Memory (KAM) Error
P0604	Internal Control Module Random Access Memory (RAM) Error
P0605	Internal Control Module Read Only Memory (ROM) Error
P0606	ECM/PCM Processor
P0607	Control Module Performance
P0608	Control Module VSS Output "A"
P0609	Control Module VSS Output "B"
P0610	Control Module Vehicle Options Error
P0611	Fuel Injector Control Module Performance
P0612	Fuel Injector Control Module Relay Control
P0613	TCM Processor
P0614	ECM/TCM Incompatible
P0615	Starter Relay Circuit
P0616	Starter Relay Circuit Low
P0617	Starter Relay Circuit High
P0618	Alternative Fuel Control Module KAM Error
P0619	Alternative Fuel Control RAM/ROM Error
P0620	Generator Control Circuit
P0621	Generator Lamp/L Terminal Circuit
P0622	Generator Field/F Terminal Circuit
P0623	Generator Lamp Control Circuit
P0624	Fuel Cap Lamp Control Circuit
P0625	Generator Field/F Terminal Circuit Low
P0626	Generator Field/F Terminal Circuit High
P0627	Fuel Pump "A" Control Circuit/Open
P0628	Fuel Pump "A" Control Circuit Low
P0629	Fuel Pump "A" Control Circuit High
P0630	VIN Not Programmed or Incompatible—ECM/PCM
P0631	VIN Not Programmed or Incompatible—TCM
P0632	Odometer Not Programmed-ECM/PCM
P0633	Immobilizer Key Not Programmed—ECM/PCM
P0634	PCM/ECM/TCM Internal Temperature Too High
P0635	Power Steering Control Circuit

DTC	Definition
P0636	Power Steering Control Circuit Low
P0637	Power Steering Control Circuit High
P0638	Throttle Actuator Control Range/Performance, Bank 1
P0639	Throttle Actuator Control Range/Performance, Bank 2
P0640	Intake Air Heater Control Circuit
P0641	Sensor Reference Voltage "A" Circuit/Open
P0642	Sensor Reference Voltage "A" Low
P0643	Sensor Reference Voltage "A" High
P0644	Driver Display Serial Communication Circuit
P0645	A/C Clutch Relay Control Circuit
P0646	A/C Clutch Relay Control Circuit Low
P0647	A/C Clutch Relay Control Circuit High
P0648	Immobilizer Lamp Control Circuit
P0649	Speed Control Lamp Control Circuit
P0650	Malfunction Indicator Lamp (MIL) Control Circuit
P0651	Sensor Reference Voltage "B" Circuit/Open
P0652	Sensor Reference Voltage "B" Circuit Low
P0653	Sensor Reference Voltage "B" Circuit High
P0654	Engine RPM Output Circuit
P0655	Engine Hot Lamp Output Control Circuit
P0656	Fuel Level Output Circuit
P0657	Actuator Supply Voltage "A" Circuit/Open
P0658	Actuator Supply Voltage "A" Circuit Low
P0659	Actuator Supply Voltage "A" Circuit High
P0660	Intake Manifold Tuning Valve Control Circuit/Open, Bank 1*
P0661	Intake Manifold Tuning Valve Control Circuit Low, Bank 1*
P0662	Intake Manifold Tuning Valve Control Circuit High, Bank 1*
P0663	Intake Manifold Tuning Valve Control Circuit/Open, Bank 2*
P0664	Intake Manifold Tuning Valve Control Circuit Low, Bank 2*
P0665	Intake Manifold Tuning Valve Control Circuit High, Bank 2*
P0666	PCM/ECM/TCM Internal Temperature Sensor Circuit
P0667	PCM/ECM/TCM Internal Temperature Sensor Range/Performance
P0668	PCM/ECM/TCM Internal Temperature Sensor Circuit Low
P0669	PCM/ECM/TCM Internal Temperature Sensor Circuit High
P0670	Glow Plug Module Control Circuit
P0671	Cylinder 1 Glow Plug Circuit
P0672	Cylinder 2 Glow Plug Circuit
P0673	Cylinder 3 Glow Plug Circuit
P0674	Cylinder 4 Glow Plug Circuit
P0675	Cylinder 5 Glow Plug Circuit
P0676	Cylinder 6 Glow Plug Circuit
P0677	Cylinder 7 Glow Plug Circuit
P0678	Cylinder 8 Glow Plug Circuit
P0679	Cylinder 9 Glow Plug Circuit
P0680	Cylinder 10 Glow Plug Circuit
P0681	Cylinder 11 Glow Plug Circuit
P0682	Cylinder 12 Glow Plug Circuit
P0683	Glow Plug Control Module to PCM Communication Circuit
P0684	Glow Plug Control Module to PCM Communication Circuit Range/Performance

DTC	Definition
P0685	ECM/PCM Power Relay Control Circuit/Open
P0686	ECM/PCM Power Relay Control Circuit Low
P0687	ECM/PCM Power Relay Control Circuit High
P0688	ECM/PCM Power Relay Sense Circuit/Open
P0689	ECM/PCM Power Relay Sense Circuit Low
P0690	ECM/PCM Power Relay Sense Circuit High
P0691	Fan 1 Control Circuit Low
P0692	Fan 1 Control Circuit High
P0693	Fan 2 Control Circuit Low
P0694	Fan 2 Control Circuit High
P0695	Fan 3 Control Circuit Low
P0696	Fan 3 Control Circuit High
P0697	Sensor Reference Voltage "C" Circuit/Open
P0698	Sensor Reference Voltage "C" Circuit Low
P0699	Sensor Reference Voltage "C" Circuit High

**DTC Application note for intake manifold tuning valves and intake manifold runner controls: Active controls are used to modify or control airflow within the engine air intake system. These controls may be used to enhance or modify in-cylinder airflow motion (charge motion), modify the airflow dynamics (manifold tuning) within the intake manifold, or both.*
Devices that control charge motion are commonly called intake manifold runner control, swirl control valve, and charge motion control valve. The SAE-recommended term for any device that controls charge motion is "intake manifold runner control" (IMRC).
Devices that control manifold dynamics or manifold tuning are commonly called intake manifold tuning valve, long/short runner control, or intake manifold communication control. The SAE-recommended term for any device that controls manifold tuning is "intake manifold tuning" (IMT) valve.

INDEX

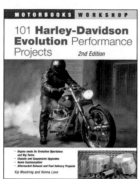

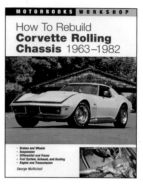